New Beginnings

New Beginnings

Why Change Is Hard and How We Can Achieve It

Stefan Klein
Translated by David Shaw

SCRIBE
Melbourne | London | Minneapolis

Scribe Publications
18–20 Edward St, Brunswick, Victoria 3056, Australia
2 John St, Clerkenwell, London, WC1N 2ES, United Kingdom
3754 Pleasant Ave, Suite 223w, Minneapolis, Minnesota 55409, USA

Originally published as "AUFBRUCH: Warum Veränderung so schwer fällt und wie sie gelingt"
Published by Scribe 2026

Copyright © S. Fischer Verlag GmbH, Frankfurt am Main 2025
Translation copyright © David Shaw 2026

All rights reserved. The publisher expressly prohibits the use of this book in connection with the development of any software program, including, without limitation, training a machine-learning or generative artificial intelligence (AI) system. Without limiting the rights under copyright reserved above, no part of this publication may be reproduced, stored in or introduced into a retrieval system, or transmitted, in any form or by any means without the prior written permission of the publisher.

The moral rights of the author and translator have been asserted.

Typeset in 12/17 pt Adobe Caslon Pro by J&M Typesetting

Printed and bound in the UK by CPI Group (UK) Ltd, Croydon CR0 4YY

Scribe is committed to the sustainable use of natural resources and the use of paper products made responsibly from those resources.

978 1 761381 68 3 (Australian edition)
978 1 917189 29 3 (UK edition)
978 1 761386 45 9 (ebook)

Catalogue records for this book are available from the National Library of Australia and the British Library.

The translation of this book was supported by a grant from the Goethe-Institut

scribepublications.com.au
scribepublications.co.uk
scribepublications.com

EU safety information: Easy Access System Europe, Mustamäe tee 50, 10621, Tallin, Estonia, gspr.requests@easproject.com.

CONTENTS

Part I: Split Consciousness

On top of the volcano   3

A paralysed society   15

Part II: Seven Illusions About Progress

First Illusion: We are realists   41

Second Illusion: We love novelty   58

Third Illusion: It's always turned out fine before   80

Fourth Illusion: Knowledge is power   99

Fifth Illusion: Freedom is the answer to everything   118

Sixth Illusion: We only want the best   142

Seventh Illusion: Ideologies are obsolete   163

Part III: How Change Works

Escaping addiction   189

The great liberation   215

An instruction manual for a better world   241

Notes   267

Bibliography   281

Part I
Split Consciousness

On top of the volcano

'To see what is right and not do it is the worst cowardice' is a maxim ascribed to Confucius. But that's easier said than done. I have never experienced the absurdity and the consequences of human obstinacy more powerfully than I did while visiting Montserrat. The tiny Caribbean island has everything required to deserve the name 'paradise'. The sea sparkles deep blue and crystal clear, as palms bend over deserted beaches. Friendly people live among lush vegetation, and rare, bright yellow orioles twitter amid the foliage. For decades, Montserrat was a favourite among rock stars. From the Rolling Stones to Sting, and from Michael Jackson to Elton John, they all liked to withdraw to the island's studios to record their albums.

And Montserrat also has a volcano. That was the reason for my visit to the island in June 1997. My editor had sent me there to write a report about volcanology. What better subject than the thousand-metre-high Soufrière Hills, described by geologists as the most active volcano in the world? I was also planning to write about the lives of those who made their homes in the shadow of the lava dome, which could explode at any moment. An eruption two years earlier had buried the island's main town, Plymouth. Its residents were evacuated in good time.

The photographer I was travelling with wanted to take some shots of the so-called ghost town. So, having obtained the relevant permits, off we set. A huge cloud of stinking, sulphurous dust rose up in the wake of our jeep, and the closer we came to Plymouth, the more we began to cough. The toxic volcanic ash was irritating our lungs. On arriving, we were met by an apocalyptic scene: the streets were buried beneath masses of lava, and most of the houses had collapsed.

We believed we were the only ones there. Plymouth had not only been reduced to little more than a toxic wasteland, but it was also in the direct firing line of the volcano, which was threatening to erupt again at any moment, and the government was allowing access only for a few hours at a time. But then we spotted people among the ruins. An old man was cooking his lunch on an oil drum, while a young woman was washing an older woman in the entrance to a collapsed house. On the barren expanse that must once have been the town's main square, two dozen men and women stood chatting among the ashes, while a couple of children played in the toxic dust on the ground at their feet.

They took almost no notice of us. What were they doing here? They didn't look as if they had come to the exclusion zone for a quick visit. I asked a woman as she unlocked the door to a derelict building. 'We live here,' she replied, inviting us to come in. She led us into a room, gloomily lit through a broken window. The space was furnished with a narrow double bed, a small table, and two chairs. Everything was covered in ash. I asked her why she had not gone to live in the new accommodation provided by the government in the north of the island, where it was safe. 'This is our home,' she replied as she put some water on to boil for tea. 'I was born here. I've always lived here. And I intend to die here.'

Dogged and undaunted

Two days later, the volcano erupted without warning. This eruption was far more powerful than any that had gone before. A column of smoke rose several kilometres into the sky above Soufrière, as rivers of lava belched from the crater and hurtled into the valley below. Helicopters were sent to save the people who still remained in the exclusion zone. And, as pyroclastic flows — deadly mixtures of hot gas, rocks, and ash — bore down on Plymouth, Royal Navy ships appeared in the waters off the small town, since Montserrat is a British Overseas Territory. The marines took those still holding out in the town on board their ships.

But, days later, when all those missing had been found and moved to the safe northern part of the island, the military helicopters still buzzed in the skies, searching from the air for people who had started heading back after they were rescued. They trudged up the mountainside, over hot ashes along the old mule tracks back to their houses and sweet potato fields on the slopes of the volcano. They wanted to return to the land that had been given to their forefathers when they were liberated from slavery. They were unwilling to abandon that land, despite the fact that it was now covered in metres of glowing-hot scoria. Confucius certainly could not accuse them of cowardice for taking no action.

And then, at the helicopter landing pad, we observed, the photographer and I, what those who had been rescued for a second time did after clambering out of the aircraft. As soon they were able to escape the attention of the military in the general chaos, they set off once again for Plymouth and the villages where the heat of the volcano had even melted houses made of concrete. After the rescue helicopters dropped off their passengers, they immediately took off to gather up the dogged islanders once again. However, the number of body bags unloaded from the helicopters increased with every flight.

Paralysed by fear

Of course, we're left shaking our heads when people risk their lives with little or no hope of success, simply because they are unable to adapt to altered circumstances. We also feel sorry for such people, who are so desperate that they will defy all reason and risk everything to return to their prior existence.[1]

But is the behaviour of the farmers in Montserrat really so alien to us? We are neither in post-apocalyptic shock, nor are we in danger of losing our minds from pain or grief. However, we do experience less dramatic forms of this denial of reality on a daily basis. It might be an acquaintance who puts up with a stressful and unsatisfying job for years although she has no shortage of better opportunities, or it might be a friend who accepts constant humiliation from his partner in the vain hope that the feelings of the honeymoon period in the relationship might someday return. And which of us has not caught ourselves summarily rejecting excellent advice about how to change our lives for the better? We know exactly what will happen if we don't finally start taking more exercise or drinking less alcohol. We make good resolutions, make tentative attempts to keep them, and soon fall back into our old habits. And then we console ourselves by believing the consequences surely won't be all that bad.

Groups of people have an even stronger tendency to shut out reality than individuals. Anyone with experience of the internal workings of an organisation will be familiar with this. One of the most difficult challenges for any company, public body, or sports club is to move with the times. There is an entire consultancy industry built around 'change management', which offers ways to make new programmes appealing to reluctant employees, club members, or customers. Most of those expensive consultancy fees are wasted, however, because any project is almost certainly doomed to failure as soon as there is even a hint of changing the prevailing circumstances. According to a survey of thousands of top business executives around the world, more than 70 per cent of initiatives that are dependent on change will fail, no matter the nature of the intended change.[2]

It should come as no surprise, then, that many people have mixed feelings when they realise that the world does not stand still at all. We see the signs all the time: trees shedding their leaves even before the summer is out; our children using chatbots to answer their homework questions; and the increasing number of old people, along with the decreasing number of young people, on our streets.

We know that, in the coming years, we will have to question almost everything we currently take for granted. We recognise that the climate crisis, the rise of artificial intelligence, the alarming rate at which our society is ageing, and our own ageing process can only be managed by striking radical new paths. We are aware that an Earth on which eight billion people enjoy the kind of prosperity we have come to take for granted cannot exist. We fear that the very survival of the human race may be at stake. We are, after all, surrounded by 'time bombs with fuses of less than 50 years', as the American evolutionary biologist Jared Diamond put it. So it is far from an exaggeration to compare our world to the volcano on the island of Montserrat.

And yet we resist change. At a time when every individual and humanity as a whole need to embrace change, we are paralysed by the tyranny of habit.

Take, for example, the schizophrenic attitude we currently have to climate change. We are spooked by storms and devastating floods, and no reasonable person doubts that entire swathes of some countries will soon be transformed into deserts if we continue to burn oil and coal. Far more people already die around the world every year from the effects of the particulate matter released by fossil fuels than were killed by the entire Covid-19 pandemic.[3] There is also no doubt that infectious diseases will spread more easily in Germany and around the world due to climate change.[4] And we don't even like petrol, kerosine, oil, and gas. We complain on a daily basis about the problems they cause: noise, pollution, traffic congestion, sky-high heating costs, long lines at the airport, and stifling heat in the summer.

We might think that this would make it easy for us to say goodbye to all that. There are alternatives, after all. But, on the contrary, the more horror stories we hear about climate collapse, and the more we feel the negative effects of our habits directly ourselves, the more fiercely we defend our cars, our right to fly to our holiday destinations, and even the central-heating boiler in the basement. We refuse even to consider the possibility that living a different way might actually make us happier.

We know full well that we cannot stop the coming changes. And so we don't even claim that that is what we want. Our plans are far more modest: new paths, yes, but not just now, please. In a world of uncertainty, we look to our old habits for stability. We cling to them like drowning passengers grasping on to the last bit of flotsam from a sinking ship.

The chapters that follow include reports of the way we become frozen with fear in the face of issues like immigration, the ubiquity of ever-more powerful computers, and the demographic ageing of our society. It is not only as a society that we are petrified with fear; the paralysis affects every individual, too. We observe this phenomenon in the people around us and, if we are honest, in ourselves, as well: when we see a chance and we know it would be a good idea to grasp it, we often don't take the opportunity, and everything remains the way it was.

Excuses are easily found. As always, it's down to a lack of money. If we ask why society doesn't change for the better, we are told it's the fault of incompetent politicians, corrupt elites, and the shortcomings of capitalism. But those explanations sound strangely half-hearted. On the one hand, no one can dispute the importance of such factors. But, on the other hand, everyone admits that they are not the only reasons for, or even the main obstacles to, positive change. We know there is something more. Why else would we hear so much about 'change fatigue' and 'separation anxiety'? Terms like these are simply ways of explaining the fact that the reason change fails is not the prevailing circumstance, but those it would affect—people don't want it. That is the most important thing to realise. Change is not just a social problem; it is also, and most importantly, a cognitive one.

But why is this so? We often hear that we show resistance to change when we are overwhelmed by the situation we are currently in. That shifts the blame back to 'the circumstances'. The common explanations of 'change fatigue' and 'separation anxiety' are rarely completely misplaced. But they do not look below the surface, and they obfuscate the issue more than clarifying it. And, most problematically, they trivialise the problem. They accept deadlock, and even denial of reality, as consequencse of a regrettably negative situation, as if they were inevitable.

Deadlock is not inevitable

This book takes the resistance to change seriously. It considers it less as a consequence of specific circumstances than as a result of our profoundly contradictory nature. On the one hand, curiosity is one of human beings' strongest natural urges; on the other hand, we instinctively reject anything new. At one and the same time, we yearn for a better life and want to keep our private world unchanged.

Reluctance to accept change is not peculiar to our time. Previous generations also rejected the new, or at least viewed it with mixed feelings. But this is the first time in human history that the very basis of our global existence is under threat. We no longer have the luxury of asking how much change we might accept. Change is coming, whether we want it or not.

Blaming the circumstances or bemoaning humanity's lack of common sense will not help. Similarly, any efforts to counter the general indolence with well-meaning words, warnings, or coercion are doomed to fail. As long as such exhortations ignore our human nature, they are more likely to increase resistance than break it down. The experience of the environmental movement, of anyone who advocates for a fairer world, or even of anyone who has simply tried to take their own organisation in a new direction is bitter. Most fail due to a false understanding of our willingness to accept change.

Yet societies are indeed capable of changing and starting anew. No one predicted the fall of the Berlin Wall even a year before it happened. Again and again, people have cast off millennia-old customs and begun living differently—from the abolition of slavery to the unification of Europe, gender equality, and equality for people of different sexual orientations. None of those revolutions is yet complete. But all have achieved in a very short time something that was previously unthinkable.

Humans are even capable of worldwide cooperation. A joint effort among all the world's nations succeeded in suppressing the AIDS epidemic and eradicating one of the deadliest diseases—smallpox—forever. The Montreal Protocol saved the Earth's ozone layer from the brink of collapse in the nineteen-eighties. These success stories are easily forgotten because we already take them for granted. That is a mistake.

How can we achieve the breakthrough we now need? We are obstructing ourselves. The only way ahead, therefore, is to gain a better understanding of ourselves and of others. Change requires that we learn to listen to each other and to take each other seriously. Can we uncover the fears that are holding us and everyone else back, identify the illusions that enfeeble us, and recognise how we allow ourselves to be manipulated?

We can only avoid falling into the traps we lay for ourselves if we know how to recognise them. Our wellbeing is now at stake, as is that of future generations. In recent years, science has shown that the illusions and bad decisions that jeopardise our future are the same as those that are already negatively affecting our sense of general satisfaction with life now. But science has also shown us ways to escape that future. It is high time we made use of those insights.

Our future depends on our ability to learn to embrace the new. Shipwrecked passengers can be rescued, as long as they manage to keep their heads above the water.

A paralysed society

Lost civilisations fascinate us because we sense that we could face a similar fate. We read chronicles of their history as if they were medical records. When we read of a sick patient's symptoms, we can't help wondering whether we're showing them, too. And if we do recognise those signs, might it not be possible, perhaps even probable, that our cities will soon be shrouded in the kind of deadly silence that so often spooked the discoverers of long-dead cultures?

> We ascended by large stone steps, in some places perfect, and in others thrown down by trees which had grown up between the crevices ... Our guide cleared a way with his machete ... [and] conducted us through the thick forest, among half-buried fragments, to fourteen monuments: ... one displaced from its pedestal by enormous roots, another locked in the close embrace of branches of trees; another hurled to the ground,

and bound down by huge vines and creepers; and one standing, with its altar before it, in a grove of trees which grew around it, seemingly to shade and shroud it as a sacred thing … The only sounds that disturbed the quiet of this buried city were the noise of monkeys moving among the tops of the trees and the cracking of dry branches broken by their weight.[1]

This is how the American explorer John Stephens described part of an expedition to the remote mountains of western Honduras in 1839. He had suspected that was where the ruins were, thanks to an old account written by a Spanish conquistador. But what he discovered in a valley a few miles from the Guatemalan border took his breath away. Beyond the Copán River, huge walls that were once part of a fortification rose out of the jungle to a height of 30 metres. Stephens and his companions crossed the river and climbed the ruins. Stretching out before them they saw a labyrinth of walls, overgrown by huge mahogany trees. Clearly, no people had lived there for many centuries. Stephens was able to make out the remains of gates and temples, and a stadium and a palace, all weathered by the tropical rains. Above it all rose the apex of a pyramid. Next to the pyramid, an apparently endless set of steps rose seemingly aimlessly towards the heavens, engraved with hundreds of hieroglyphs.

Who had built this city, and why had they abandoned it? The ruins of Copán are located in the ancient territory of the Maya, and Stephens knew of that advanced civilisation's achievements. Mayan architects were no less skilled than their counterparts in the Old World. Mayan farmers cultivated maize in cleverly terraced and irrigated fields. Mayan astronomers working in specially constructed observatories predicted the paths of the stars, following not one, but three different and highly complex calendars. Mayan mathematicians used a number system that included the concept of zero, centuries before Europe abandoned the cumbersome Roman numeral system. They had writing, literature, and libraries. Their bark-based *amate* paper was far superior to European parchment.

Something — or someone — must have wiped out this great civilisation. It was not the raiding Europeans who brought about its downfall. This is what distinguishes the Maya from many other Native American nations: their empire was already in ruins when the European conquerors arrived. Even the account by the Spanish pioneer that had led Stephens to explore the jungle in the first place told of overgrown ruins on the banks of the Copán River.

Stephens supposed an unknown disaster must have led to its collapse: a period of drought or an earthquake, an epidemic, or a devastating war. This civilisation existed for hundreds, if not thousands, of years until reaching its peak. The idea that such a culture could have brought about its own downfall was unthinkable to Stephens.

The anatomy of a downfall

We now know that Stephens was partly right in his assumption, and partly wrong. Copán was a Mayan city, and the Mayan civilisation was among the most enduring in human history. Its beginnings date back to the third millennium before the Common Era. Mayan cities began to grow towards the end of the second millennium BCE, and the Maya developed writing and a state system during the first millennium. Their civilisation flourished especially during the five centuries prior to its sudden demise in the 9th century CE. Thus, the civilisation endured for longer even than Ancient Egypt did.

It really was a disaster that brought about its downfall. But it was neither conquerors nor earthquakes that spelled the end for the Maya. The mystery was solved by a stalagmite taken from cave known as 'The Palace of Kings' in the Mayan language. In a series of groundbreaking publications from 2010 to 2020, geologists reported the results of radiochemical analyses of the stalagmite, as well as drill cores from the bottom of several Central American lakes.[2] The sediments contained in the stalagmite and the drill cores provided scientists with precise information about temperature and humidity conditions in the past. The results consistently indicated a virtual climate collapse in Central America around 800 CE.[3] This was also the period when the Mayan civilisation collapsed.

Average rainfall dropped by a half, in some cases even plummeting to a quarter of its annual level. Geoscientists call such periods of extreme dryness 'megadroughts'.[4] Since time immemorial, the rainy season in Mayan lands was predominately in the summer, and the irrigation systems for the Maya's ever-more intensive agriculture collected the water from those downpours and stored it to survive the dry winter. However, those vital heavy summer rains began to fail. The great water reservoirs built by the Maya began to dry up.[5] Millions of people died of thirst and starvation, and wars and unrest broke out.

However, the Maya were not simply victims of the forces of nature; far from it. Contrary to Stephens' assumption, their own actions were also to blame. Current research supports the theory that the Maya themselves triggered the climate change that wiped out their civilisation. Although there had always been naturally triggered, protracted periods of drought in Central America, the turn of the first millennium saw less precipitation in the land of the Maya than had ever been the case since humans settled the region.

The megadrought began at a time when the cities of

the Maya had grown more magnificent than ever, their rulers wielded unprecedented power, and their culture reached its peak. This was no coincidence, according to the American climatologist Benjamin Cook and his colleagues.[6] As Mayan culture and technology developed, the population grew. The people cleared the forests to use the land to grow increasing amounts of maize. This firstly caused soil erosion, since maize grows in broadly spaced rows and has such shallow roots that they cannot prevent rain and wind from carrying away the soil. The land began to dry out. The Maya had to clear more and more forest for their maize fields. The deforestation had a fatal effect: because less water was able to evaporate from the soil in the agricultural fields than could be released by the dense foliage of the forest, the air humidity fell. This, in turn, led to less rainfall.

A vicious cycle had been set in motion. The deforestation caused a 20 per cent drop in summer rainfall—precipitation that was relied upon for agricultural production. In good years, when the rain god was merciful, there may have been just enough water for farming. But this man-made climate change had robbed the environment of its natural reserves. Even the slightest meteorological disturbance was now enough to bring about the collapse of agricultural production. Famine was followed by anarchy and war, and presumably also by disease. Mayan civilisation had reached a tipping point from which it was not able to return to stability.

The ruling houses of Copán, however, do not seem to have been inordinately concerned. The Mayan rulers were mainly engaged in competitive temple-building, trying to outdo each other with ever-larger edifices, which the American geographer Jared Diamond has described as 'reminiscent in turn of the extravagant conspicuous consumption by modern American CEOs'. Did they believe their world would exist forever? It seems that no one was willing, or able, to imagine such a magnificent, millennia-old civilisation ever perishing. And barely a generation later, it happened. Around the year 850, Copán's royal palace burned down, never to be rebuilt. The few survivors of famine, drought, disease, and killing left the cities. The population, now a fraction of its former size, fled to places where there was still water all year round, and attempted to grow what maize they could there. The expansive fields that had once fed the cities were reclaimed by the wilderness, as trees began to take root among the ruins. And, as the forests expanded again, the summer cloudbursts also returned. But the civilisation of the Maya was lost forever.

For they knew not what they did

It is tempting to compare the arrogance of the Maya with our own. However, hastily drawn parallels overlook one vital difference: the Maya did not know what they were doing.

There is no indication that they knew of the complex interplay of the way that trees reflect sunlight, or absorb humidity and affect seasonal precipitation patterns, as discovered by 21st-century scientists with their supercomputers. For the Maya, droughts were caused by the whims of Chaac, the god of rain, thunder, and fertility. And the only way to change them was to appease him. They made offerings of sacred maize, and slain jaguar, or even human sacrifices, to implore Chaac to swing his lightning axe and shatter the clouds to unleash the rains.

Many ancient civilisations have brought about their own downfall with their actions. None of those societies was able to connect the dots. The Indus Valley civilisation in modern Pakistan, for example, was one of the three earliest advanced cultures in history. It was probably the first civilisation to perish due to an environmental disaster of its own making. The collapse came at the turn of the second millennium BCE. Scholars in the magnificent cities of the Indus Valley, where there were even flushing toilets, had no idea why the crops were failing. We now know that the reasons were similar to those responsible for the collapse of the Mayan civilisation. Overcultivation caused soil erosion; deforestation caused a drop in rainfall. When climate change then disturbed the cycle of monsoon rains, agriculture collapsed, and, with it, the cities.[7]

Even the historiographers in Ancient Rome failed to

understand what was happening around them. Important historians of Late Antiquity such as Ammianus Marcellinus, while they recognised that the days of the Empire were numbered, put the blame on corruption, the decline of traditional morals, and barbarian invasions. They failed to see that those were the symptoms rather than the causes of the unravelling of power. Today we know that neither decadence nor horsemen from the east were the cause of the Empire's demise.

In fact, the Roman Empire was a victim of its own success.[8] Rome's occupation of more and more territory became so costly that, from the 3rd century CE onwards, it was no longer able fund it from the taxes it collected. This was compounded by the fact that Rome's economic system relied on pillaging the provinces. Roman engineers built masterful constructions and brilliant machines of war. But none of them seems to have been interested in improving industrial technologies or increasing agricultural productivity. The fields of Imperial Rome were farmed using largely the same primitive tools as those used at the time of Alexander the Great.[9] The large landowners had slaves to do the work, after all. If they had too few workers, the land would lie fallow, further reducing the state's tax income.[10] And so the Empire was cursed by the very customs that had made it great. The Empire was doomed to die because its society was incapable of change.

The decline of the Roman Empire, famine in the Indus Valley, the fall of the Maya, and the collapse of many other civilisations all have one thing in common: the people were blind to the disaster they were hurtling towards. They could not avert the disaster because they understood neither the situation they were in, nor their role in creating it.

Hurtling towards collapse with eyes wide open

We, on the other hand, are well aware of the situation we are in. We see in the media every day just how we are doing. We are confronted with images of devastating floods, dried-up riverbeds, and burning forests on every continent. We read that 150 plant or animal species disappear forever every day, simply wiped out. And we ultimately learn that my own country, Germany, has no right whatsoever to claim to be top of the class. Germany emits more carbon dioxide per capita than any other country in Europe.[11] If it is to do its fair share to limit global warming to 1.5 degrees, Germany must reduce its greenhouse gas emissions 30 times faster than it has done so far.[12]

Experts tell us this is just the beginning, since temperatures have only now begun to rise and the 'great dying' of the natural world will only continue to pick up pace. In just a few years' time, rising sea levels will flood our port cities while our fields grow increasingly arid and entire countries become uninhabitable.

Do we need any more wake-up calls from the climatologists? We can see for ourselves what is happening. Our lawns become brown earlier every summer, and frozen lakes in winter are little more than a memory. You would have to be living in a cave with all your senses shut off to keep believing we live in an intact world. The global climate crisis is an emergency on an unprecedented scale in human history.

Machines out of control

As if that weren't enough, the destruction of the basis for our way of life is far from the only danger we knowingly find ourselves in. Computers have transformed almost every aspect of our lives in a very short period of time. As screens control our work, algorithms control our opinions, and dating apps control our most intimate lives, it is already difficult to say whether we control machines or they control us. And here, too, we are only at the beginning. Compared to the technology we will be using in just a few years' time, today's computers will seem like harmless children's toys.

Artificial intelligence undoubtedly offers many opportunities. It will free office workers from tedious, mundane tasks, assist doctors in making diagnoses, and help scientists find new medicines. It will tailor learning plans to individual school students and optimise psychological treatments. Driverless cars and workerless factory floors will be the norm in a few years' time. Per-capita economic output could increase tenfold when artificial intelligence reaches the capabilities of the human mind, according to the United Nations.

This technological progress is also changing society. An open letter signed in March 2023 by more than 100 of the world's leading researchers in the field of artificial intelligence describes what lies ahead.[13] It points out that AI systems are increasingly able to compete with humans, and it goes on:

> We must ask ourselves: Should we let machines flood our information channels with propaganda and untruth? Should we automate away all the jobs, including the fulfilling ones? Should we develop nonhuman minds that might eventually outnumber, outsmart, obsolete and replace us? Should we risk loss of control of our civilisation?

There is no better way to put it. However, this list is far from complete. For example, the letter fails to address the fact that artificial intelligence is also an excellent surveillance tool.

Even if we succeed in preventing such excesses, machines in the future will do more than simply carry out orders. They will take over the work of business executives, lawyers, and doctors, including decision-making. Conservative estimates put the number of jobs lost in this way in industrialised nations over the coming years at 300 million. Two-thirds of all jobs are likely to change in some way due to the use of artificial intelligence.[14]

The United Nations has warned that the enormous wealth promised by this new technology is likely to increase the tensions that are already being felt in almost every country. It predicts that the huge profits are likely to be concentrated in the hands of a few, and that many will be left behind. The UN also warns that concentrated technological and economic power translates into political power, and leads to a powerless majority in a state of discontent.

While the climate crisis poses an external threat to our civilisation, the uncontrolled computerisation of society could split it from the inside. Growing inequality and polarisation threaten to tear it apart. This does not mean that Western democracies are heading inexorably into the abyss. The challenges posed by ever-faster technological progress can be overcome.[15] But to believe that our society can carry on as before with impunity is dangerously naive.

When asked which possible impact of the social media already at large in our society he found most worrying, the former Facebook executive Tim Kendall answered with just two words: civil war.[16]

The great ageing

And, finally, there is a third development forcing a change in the way we have become accustomed to living. It is slower than climate change and the rapid rise of artificial intelligence, but we can already feel its effects. When we go out onto our streets, we see more and more older faces, and there are fewer people to serve us when we meet at a café.

This development can be summed up in a single number: 31. That was the number of pensioners for every 100 people of working age in Germany in 2020. This so-called 'old-age dependency ratio' stood at 20 in the year 2000. That means it rose by more than half in the space of just 20 years, and it is set to have doubled to 47 by 2040. Every retiree's pension will be paid for by two workers.[17] These figures describe Germany's future, but the picture is similar in all developed countries. Even most parts of Africa and Asia are registering ever-decreasing birth rates. Humanity is ageing all around the world.

That more and more people are living longer and longer lives is indeed good news. The problem is that our society is not designed for a situation in which people over 50 almost make up the majority, which is already the case in Germany. The inconvenience of having to wait weeks for a repairwoman or handyman to come around, because there is simply no one to do the work any sooner, is just a foretaste of times to come.

With the coffers empty, we face a future in which nurses and doctors are in short supply. And those are just the most obvious effects of a drastically ageing world.

Most people barely notice the most momentous changes, as they experience them happening gradually within themselves. It is not only our bodies that age as we get older; our minds do, too. It is not the case that older brains are less capable than younger ones, but they work differently. Developmental psychologists have described these processes in detail. Older people take fewer risks and are generally more fearful. They trust in their experience and are less likely to try new ways of solving problems. Their drive to experiment becomes less pronounced.

Between the ages of 20 and 40, Pablo Picasso invented no fewer than five revolutionary painting styles. He went through his Blue Period, his Pink Period, Analytic Cubism, and Synthetic Cubism before finally developing his own form of surrealism. As an old man, Picasso was still extremely productive, remaining a masterful painter until his death, but he repeatedly quoted himself in thousands of works.

Less extraordinary minds age in exactly the same way. An analysis of patent applications from more than one million American inventors proved that their ideas changed over the years.[18] The younger the inventors were, the more likely they were to patent an invention that later turned out to be groundbreaking. The older the inventors were, the more likely their patent was to be for a sequential improvement on some existing technology.

The strategies employed by older people are not worse than those of younger people; they are simply different. A society that is dominated by old people approaches challenges in a different way from a young society. There is a tendency to stick with the familiar rather than to forge new paths.

'A new scientific truth does not triumph by convincing its opponents and making them see the light, but rather because its opponents eventually die.' The physicist Max Planck, a conservative man, knew what he was talking about. His principle applies not only to science, but to everything else, too. The younger people are, the more they are able to adapt to altered circumstances. In a stable environment, the caution and experience of older people is advantageous. But in a rapidly changing world, an overly conservative society runs the risk of decline due to its inability to react quickly to altered circumstances. And that is the kind of world we live in today.

Two souls, alas …

We must adapt. That much is certain. It puts us one step ahead of the civilisations that collapsed because they failed to see the writing on the wall. But how much willingness to act does our knowledge provoke? Eighty-nine per cent of people in Germany, for example, consider the climate crisis to be a 'serious' or 'very serious' problem. Eighty-seven per cent want greenhouse gas emissions to be reduced as quickly as possible, so that Europe becomes climate neutral by no later than 2050. Seventy-six per cent of Germans claim to have taken action to fight climate change themselves in the previous six months.

All three statistics come from the 2023 *Eurobarometer* survey for the European Commission.[19] Such high approval ratings are more than a momentary snapshot. They have been consistently high for years. Before Russia's invasion of Ukraine, a huge 92 per cent of those questioned considered the climate crisis to be a 'serious' or 'very serious' problem. Concern for the planet is very close to German people's hearts.

It might be expected from these figures that efforts to save the environment and the climate would be met with enthusiasm. Even considering the fact that Germans are not exactly prone to exuberance, their level of support for environmental issues should mean audible sighs of relief could be expected when a meaningful change in legislation to protect the atmosphere is passed after years of delay.

But what happened when the German government announced in the same year, 2023, that it would stop registering cars with combustion engines after a generous transition period of more than ten years? Once again, opinion polls pointed to an overwhelming majority: 65 per cent of Germans opposed the move. They still wanted to be able to fire up a turbocharged four-cylinder engine in their next new car but one—despite their awareness of the fact that such engines pump almost half a million tonnes of greenhouse gases into Germany's air every day.

Supporters of the combustion engine claim that it is the only affordable technology for many people, but that is not the case. In the same year, the average price paid for a new car in Germany was 44,600 euros.[20] Good electric cars can also be bought for that price. And used petrol or diesel cars would also still remain available. Freely available information from Europe's largest automobile association, the German ADAC, shows that even back in 2023, electric cars cost their owners less than diesel or petrol vehicles over their entire working lives, and indicated that they would only become cheaper in the future.[21] But that voice of economic reason went unheard amid all the media clamour. Did people really think they should not or could not live without the roar of engines, the smell of petrol at the pump, and the stink of exhaust fumes?

What happened to that so fervently expressed desire for an intact environment?[22] Cars are closer to Germans' hearts than any other object of daily use. However, it would be unrealistic to dismiss the discrepancy between insight and action as nothing more than a national obsession with horsepower and metallic paint. It is a universal phenomenon. A large-scale investigation covering 20 countries concluded laconically that citizens' concerns about climate change were not indicative of their support for the corresponding action.[23]

Just think of the outrage whenever anyone suggests that more people should give up eating meat every day, not only because of the climate crisis or the conditions under which livestock is farmed, but also for the sake of their own health. Or think of people's reluctance to accept immigration from other continents, even though no one denies that Europe already faces a huge labour shortage.

Naked without a screen, naked before the world

And what about the digital revolution? What the car was to older generations, the smartphone is to the younger one. It is a very special thing. It is almost impossible to imagine life without it — an electronic device in our pockets that functions like an organ of our own bodies.

Nonetheless, no one is comfortable with this symbiosis. Detailed investigations have provided impressive proof of a long-suspected link: the more people form their worldview from social networks and other online channels, the more likely they are to distrust other people and institutions, to feel and spread hate, and to be polarised into opposing camps.[24]

However, that unease is rooted in more than just concern about the threat to social cohesion posed by digital media. Even in our ordinary, everyday lives, we have the feeling that there is something not quite right with our increasingly virtual existence. Which of us has not been shocked by how jittery we become when the internet goes down? Everyone instinctively knows it, and scientific studies have proven it countless times: we would be far more content and balanced if we spent less time in front of screens.[25] And almost everyone has tried at some point to log off the net for a few hours or days — only to realise how empty, helpless, and naked we feel without a touchscreen within reach at all times.

And what about our awareness of the fact that we are exposed on the internet against our will? Eighty per cent of people in Germany are mistrustful of the online economy.[26] On the one hand, we dislike the idea that the information about our preferences, thoughts, feelings, and character traits that we give away with every mouse-click will certainly not be used to our benefit. Only 3 per cent of those questioned in Germany said they did not care at all how Google, Amazon, etc. handle their data. But, on the other hand, we act precisely as if we cared nothing about it. Eighty per cent of people in Germany willingly make use of services they suspect of breaking data-protection laws.[27] They would rather use shady, free apps than pay for less prying alternatives. One survey among Facebook users highlighted this schizophrenic attitude. Many users expressed a fear that strangers might find out about their sexual orientation or political views. However, among the very users who were most worried about this, almost half had posted information themselves about their sexual orientation and political views.[28] Of their own free will. It seems the promise of a few likes was enough to banish all their misgivings.

The art of self-consolation

We want to live in dignity, with a secure future for ourselves and our children. At the same time, we undermine those very goals. In other words, our values are in contradiction with our actions. Sometimes they are diametrically opposed.

People have always agonised over internal conflict. More than 2,000 years ago, the Apostle Paul lamented in a famous passage from his Epistle to the Romans: 'For the good that I would do, I do not; but the evil which I would not do, that I do.' This shows an admirable degree of honesty. Paul resisted the temptation that many people succumb to when they are in conflict: the temptation to simply deny that the conflict exists. Paul deplored hypocrisy, which turns the unwanted into the wanted, and reinterprets a person's own weakness as a virtue. He wanted to be honest, even to himself, even if it was painful to do.

However, in the very next verse of his epistle, Paul surprises us with a sophisticated thought. He claims that the author of his misery is not himself, but rather an external force that cast him into misfortune. 'Now if I do that which I would not do, it is no more I that do it, but sin that dwelleth in me.'

Paul therefore denies responsibility for his own actions. He declares that he is powerless in the face of sin. He hopes for an unspecified redemption that would make everything right again.

Lines of thought such as this described first by the Apostle are known as 'techniques of neutralisation'. They are particularly important in criminology: lawbreakers use techniques of neutralisation to convince the world and, most importantly, themselves that they are not actually guilty of the crimes they commit. Attentive readers will recognise Paul's strategy from every tabloid talk show on TV, as well as from many of their own inner monologues. We just use different words. Instead of 'sin', we speak of 'economic constraints' in our defence. For instance, if we cannot afford to buy something we want or even consider necessary—which, all too often, means that we give priority to other objectives. When politicians make decisions against their better judgement, they often point to the fact that they cannot 'ask too much of voters'—which is simply an expression of their fear of displeasing their electorate. Then they pledge salvation in the form of a promising but not-quite-matured new technology, or an international agreement sometime in the future.

Could we even bear to live with ourselves without such techniques of neutralisation? Saint Paul could not have imagined the contradictions that humans would eventually tangle themselves up in. The Roman Empire may have been one of the greatest and most complex in history, but compared to today, Paul lived in an uncomplicated world. People were poorly informed, consumed little, and had little or no influence on political events. Their moral behaviour was concerned only with those close to them.

The contradictions of the early 21st century, by contrast, stretch our moral sense almost to breaking point. Even our everyday decisions impact the farthest-flung parts of the world and people who have not even been born yet. And no one can plead ignorance. The connections are common knowledge. Anyone who buys a T-shirt is almost certainly supporting exploitation in Asia by doing so. And anyone who drives a car pumps a burden for future generations into the atmosphere every time they start their engine. Anyone cutting into a steak is not only aware of the misery suffered by factory-farmed animals, but also that cattle are responsible for a large share of greenhouse gas emissions, and that mass farming is a driver of deforestation.

It is no good arguing that our own little contribution is negligible. It is not. In a widely acclaimed study, the American ethicist John Nolt, for example, estimates that the amount of carbon currently emitted by an average citizen of a developed nation over the course of their life is enough to kill at least one future person.[29]

At the same time, we are moral beings. We want to be good people, and we often are. Psychologists call the state of mind first described by the Apostle Paul 'cognitive dissonance', and we suffer from it more than any humans before us. Our minds are split between desires, attitudes, and urges that appear mutually incompatible, and we are sometimes fully aware of those contradictions. Cognitive dissonance leaves us feeling the same way as listening to someone singing out of tune. Indeed, the phrase is rooted in the Latin term for 'mental discord'. The human mind craves harmony. And the most painful contradictions are those we create ourselves—for example, by engaging in behaviour that we know is harmful. This is our dilemma. It is almost enough to make us envy the ancient Maya and all the other lost civilisations, which at least knew nothing of the ruin they had inadvertently brought upon themselves.

Part II
Seven Illusions About Progress

First Illusion:
We are realists

It was purely by coincidence that Leon Festinger discovered the sophistication with which we are able to put a positive spin on the world around us. A social psychologist at Stanford University in California, Festinger was interested in what happens when prophecies fail. His interest was piqued in 1954 by a newspaper article with the headline 'Prophesy from the Planet Clarion: Flee the Flood. It'll swamp us on Dec. 21.'

The article was about a strange religious cult called the Brotherhood of the Seven Rays.[1] It developed around the mysterious knowledge of a Chicago housewife called Dorothy Martin, whom cult members knew as Sister Thedra. She claimed to be receiving telepathic messages from a planet called Clarion, and that the aliens had told her the date of the impending apocalypse. They said floods would hit America and Europe at dawn on 21 December 1954, after which a devastating earthquake would destroy the world. Only those who joined the Brotherhood beforehand would survive. The aliens had benevolently decided to send a rescue UFO to take believers to the safety of the planet Clarion.

That sounded promising. Festinger and two of his collaborators infiltrated the Brotherhood as it was preparing for this exodus. The cult members quit their jobs, sold their houses and possessions—donating the proceeds to the community—broke off friendships with people who refused to join the sect, and even divorced unbelieving partners. The Chosen invested a lot in surviving the end of the world. Everything, in fact.

During this time of preparation, they saw no reason to doubt the prophecy. The Brotherhood was well organised, and salvation seemed certain. At midday on 17 December—four days early—Sister Thedra announced that a certain 'Captain Video' had called her from interstellar space to say the UFO was on its way and would land in her back garden at 4.00 pm sharp. The Brotherhood gathered for the arrival. A gaggle of journalists who had somehow heard of the prophesied landing watched as the believers obediently cut out the zippers and any other metal from their clothing, and the women even removed their bras, in anticipation of their departure.

At four o'clock, there was no sign of a flying saucer heading their way. The Brotherhood waited and waited.

It was not until many hours later that the first members tentatively began to discuss the matter. They could easily have decided that the ominous Captain Video was really a hoax by a reporter hoping to get a good story. To assume that Sister Thedra was either quite dull witted or simply a fraudster would also have been an obvious explanation. But no one mentioned any such thoughts.

Instead, the gathering concluded that the intellectually superior aliens had called a practice drill. That would finally dispel any doubts about Sister Thedra's prophecy. The Clarionites had once again reaffirmed their utmost determination to help. The rescue was imminent. So the group dispersed, determined to do even more to prepare for the apocalypse.

This was an example of the kind of doublethink that the undercover cult member Leon Festinger later termed 'cognitive dissonance'. The Chosen knew perfectly well that there was no spaceship, and they *didn't* know it, at the same time.

A redeeming message

On 20 December, the eve of the predicted flood, Sister Thedra received a new message from space. She told the Brotherhood that the Clarionites would pick the Chosen up in cars at midnight to ferry them to the waiting spaceship. To avoid journalists on that crucial night, she called for the Brotherhood to gather at her house. Even more carefully than during the drill, the Chosen removed all metal from their clothing and sat down to wait, their overcoats folded neatly in their laps. They were ready. Nobody spoke. The only sound was the ticking of Dorothy Martin's big wall clock as it counted down the minutes to Armageddon. At one minute to midnight, when there was still no sign of any spaceship arriving, one man whispered that the clock was wrong, pointing at another clock on the mantlepiece. Its hands really did show ten to midnight. They turned the wall clock back. One minute before midnight, according to the corrected clock, Dorothy Martin declared 'the plan has never gone astray'. But it was concerning that no Clarionites had yet turned up. Someone asked if perhaps they had forgotten to remove a piece of metal. And indeed: one of the members of the Brotherhood had forgotten to mention he had a mercury filling in his tooth.

They pulled out his tooth. Still no spaceship.

Journalists called to ask what had happened to the end of the world. Sister Thedra did not take the calls. At around half past three in the morning, she was instructed telepathically to tell the group to take a coffee break. The spaceship was on its way. The mood was tense, and some of the cult members felt let down and tearful.

At 4.45, less than two hours before the prophesied apocalypse, Sister Thedra gathered the group together.

She told them that the aliens had made contact with her again. With trembling hands, she took out a piece of paper and read out the latest telepathic message, to the effect that the little group, sitting all night long, had spread so much light that God had saved the world from destruction.

That was the redeeming message that the superior beings from Planet Clarion communicated to humankind through Thedra. Would or could anyone deny it? The prophecy had been correct. God had wanted to destroy the world, but the superhuman efforts and deadly fears endured by the Brotherhood of the Seven Rays had not been for nothing. The utmost determination of the Chosen alone was to thank for the fact that the sun rose again on the morning of 21 December 1954 and that the world continued to exist. The Clarionites meant humans well. And that was the explanation Sister Thedra gave to the astonished press the next morning.

The discovery of cognitive dissonance

Festinger realised that he had been observing more than just a group of strange people who were desperate to retain their abstruse view of the world. Behind the odd ideas and bizarre behaviour of the Brotherhood of the Seven Rays was a conflict that even the most reasonable of us can find ourselves in. This conflict comes about whenever we behave in a way that is incompatible with reality or with our own beliefs, or — even worse — with both.

People often cling to decisions and habits even long after the relevant circumstances have changed. Painful discrepancies also arise when our decisions or habits prove to be incompatible with the principles we believe to be right. The divergence of habitual behaviour and perceived reality can shake societies to the core — for example, when we are told that our consumption of resources is destroying the basis for our own survival, or when a previously underestimated power suddenly starts a war. Such events foster a feeling of insecurity, just as they did among the cult members in December 1954 when they realised the world was not going to end just before Christmas.

Festinger called this state of mind 'cognitive dissonance'. Since the discord arises from a contradiction between a person's behaviour and their experience of reality, the social psychologist argued that there are two possible ways to escape the dissonance. One or the other must be changed. When no spaceship arrived, Sister Thedra and her disciples could have admitted their mistake, disbanded the Brotherhood, and returned to their suburban lives. But the cult members were not prepared to do that. So they had to come up with a new interpretation of reality.

Festinger's crucial realisation was that both options

resolve the dissonance and are, from the point of view of those affected, equally valid. Veraciousness is a moral category, but it is not relevant to the proper functioning of the brain. Brains evolved to increase organisms' chances of surviving and reproducing. They did not develop in order to acknowledge the world or to be a reliable witness to it. It is often easier to escape a conflict by sacrificing realism than by changing behaviours.

The mother of all illusions

In almost every case, the better option is to stick to the facts. We can rarely solve problems by ignoring them. Why wouldn't we confront reality head on? It doubtless leads to wiser decisions. But, instead, we take refuge in an illusory world in which the contradictions disappear. Why do we find it so difficult to admit that our behaviour is often illogical, sometimes in violation of our own principles, and can sometimes seem downright absurd? Why do we cling to decisions and habits that we know do us harm?

No one can make changes without asking themselves those questions, because resistance cannot be overcome if its origin is unknown. It is often said that people persist in their beliefs in order to maintain their sense of identity. Perhaps we cling to habits for fear of appearing inconsistent, to others and to ourselves. Fear of feeling that we are at the mercy of external circumstances, and of being seen as such, certainly plays a part. But is such a psychological theory enough to explain the extent to which we avoid reality? In fact, it doesn't explain anything. After all, when someone shows they are capable of learning by changing their mind, we could see it as a sign of strength. Our great willingness to indulge in illusions must be more deeply rooted.

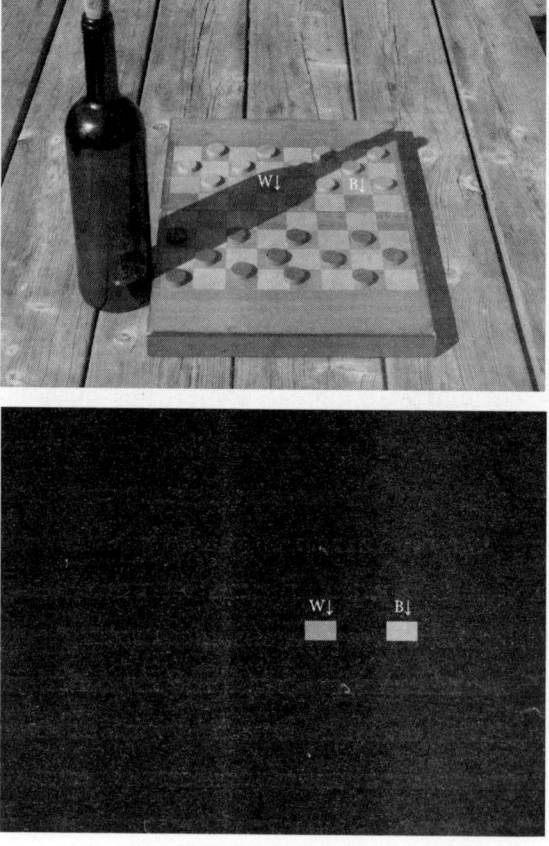

Fig. 1: Which square is darker?

Neuroscientists did not lay the foundations for a convincing explanation until around the turn of the millennium, when they gained a new understanding of the way our brains perceive reality and lead us to base decisions on that perception. According to this view, all our perceptions, feelings, thoughts, and actions are geared towards avoiding the unexpected. Our brains try constantly to reconcile reality with our expectations.

Who would not be willing to swear that the white square on the chessboard opposite marked 'W' is lighter than the black square marked 'B'? But the opposite is true. If the rest of the photo is covered, the 'W' square looks darker than the 'B' square. But when the shadow of the bottle is visible, the white square looks lighter than the black one. Our brains account for the differences in light and shade, and make the white square look to us as it would if it weren't in the shade. We are familiar with chessboards, and expect the white squares to be lighter than the black ones.

This is an example of the principle of predictive coding. This rapidly evolving theory currently provides the best explanation for our thoughts and actions.[2] 'Predictive' means that the mind works not with facts, but with prognoses. From this, it follows that our perceptions, judgements, and decisions are based on preconceptions. In the case of the chessboard, the preconception is that white squares must be lighter than black ones. Predictive coding shows that the brain creates a worldview and that it links perceptions to actions. It explains why we cling to prejudices and habits even when we have enough information to know better—and we usually don't even notice the mistake we are making. The theory also explains why individual humans and entire societies shy away from change: the brain is an illusion machine, and predictive coding is the mother of all illusions.

Figure 2 shows another example of this: at first glance, it appears to be the same mask depicted twice, and in both images it appears to bulge out towards us. But if you examine the picture on the right more closely, with the eye of a detective, so to speak, you may notice something slightly confusing. The shadows don't look right. The tip of the nose is lighter than the left side of the nose, as if it were illuminated from above. But in that case, what is casting the clear shadow on the right of the nose? And why is the temple lighter than much of the forehead?

The photo on the right actually shows the inside of the mask, and the left-hand picture shows the outside, and so they look like a mirror image of each other. The mask on the right bulges away from us, into the page, with the tip of the nose as the deepest point. We are looking into a hollow space. The shadows in the illustration are correct—the light source is illuminating both masks from the left.

First Illusion: We are realists

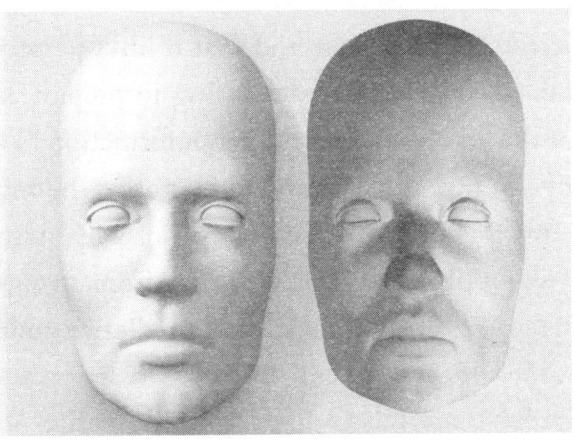

Fig. 2: Two images of the same mask?

Even after realising this, we continue to find it impossible to see the mask on the right as concave. The signals our brains receive from our eyes do not match our expectation that faces should be convex. To avoid this dissonance, our brains ignore the information conveyed by the shadows. The illusion effect becomes even more intense when we are confronted with a video animation of such a mask rotating, many of which can be found on the internet. As soon as the back of the mask appears, the figure suddenly seems to switch, and the direction of the mask's rotation appears to flip.[3]

This phenomenon is more than an illustration; it is a provocation. In Western societies, we imagine perception to work like a camera. We call it being 'objective' when we believe we are describing facts in a dispassionate and unbiased way. But even the most elementary perception shows us the world not as it is but as our preconceptions tell us it should be. Every mental activity begins with an expectation—in this case, an expectation about the shape of faces. We then attempt to see reality in the light of that expectation. This is the exact opposite of the precise detective methods employed by Sherlock Holmes. When we look at the masks, we ignore any contradictory information, even if it is irrefutable. In the case of the chessboard, our brains even falsify the information they receive from our eyes.

The world from a tin can

What we consider to be facts are actually fantasies. A few key stimuli are enough to trigger such fantasies. The pattern of a mouth, nose, and eyes, or alternating light and dark squares, generates an expectation in us. We have all seen more than enough faces and chessboards to enable us to form an image out of such an expectation.

These hallucinations usually match the information provided by our sensory organs well enough for us to be able to negotiate the world. When there is a mismatch, though, the result is cognitive dissonance. We have two options for dealing with this. The first is to adapt our expectations. Our mental image of faces and chessboards are generally in colour, but the illustrations in this book appear in black and white. When we see them, the aspect of colour is removed from the image in our mind's eye. The second time we look at the images, that perceptive step is no longer necessary. We now know what to expect, and have adapted our expectations. This is how we learn.

Alternatively, we can resolve the dissonance by denying that the mismatch exists. This is what you just did when viewing the masks and the chessboard. All too often, we are so good at disregarding confusing information that we don't even notice we are doing it. We see an illusion, and believe all is good with the world.

The fact that we continuously confuse facts with expectations does not mean our minds work badly. On the contrary: it is the only way our brains can fulfil their function of ensuring our survival and propagation as organisms. On the one hand, our brains work slowly. Even the simplest of tasks, such as recognising a flashing light, takes a few tenths of a second. More complex tasks can tax our mental capacities for much longer. This alone is incentive enough to turn a blind eye to reality when it fails to match our expectations. Clearing up contradictions always takes an effort, and does not necessarily bring any benefits.

On the other hand, perceiving, feeling, and thinking all take energy. Although the brain makes up only 2 per cent of a human's body weight, it places more demands on our metabolism than any other organ. It accounts for a good 20 per cent of the energy our bodies consume.

Our mind makes up for its slowness and sluggishness by using predictions. Predictive coding allows the brain to get ahead of its environment. Additionally, the brain is very efficient. Rather than having to process every new detail of an event as fresh information, it makes use of preconceptions. The world we experience comes out of a tin can.

Our brains live in the future

We live in a constant state of 'What if …?' It is not only our perception that is based on expectations, but also our actions. Before we do anything, our brains draw up a prognosis.

Most of those predictions sound simple. (If you're hungry, a trip to the kitchen will help.) Other prognoses are more complex. (A holiday with our partner will rekindle the flame of attraction. Taking part in a demonstration against European agricultural policy could save rare insects from extinction. Joining a brotherhood and escaping to a distant planet before the apocalypse will save your life.)

Such prognoses contain both an expectation about our environment and an assessment of how we are likely to fare in that environment. (If we eat food out of the fridge, our hunger will appease.) We consider various alternatives, and select those we deem both desirable and achievable. (Guzzling that sweet pudding you find in the fridge might satisfy our hunger more quickly and more tastily than a stick of celery from the salad drawer.)

When the brain makes a prediction it considers to be desirable, it strives to fulfil that expectation. Our actions are then aimed at delivering our organism to the predicted future state. (A full stomach and an aftertaste of chocolate on the tongue.) Once again, the brain's priority is not to identify what is real. Instead, its goal is to bring reality into line with its own expectations — to shape reality the way we want it to be. That works best in a predictable world. And that explains humans' mistrust of the new. Change threatens the validity of our predictions.

The elephant and the bull

Ludwig Wittgenstein was a pioneer when it comes to thinking about the connections between perception, thought, and action, long before the rise of modern brain science. Perhaps the most influential philosopher of the 20th century, Wittgenstein published his *Philosophical Investigations* in 1953, in which he wrote about ambiguous images.[4] He used an image popular at the time of a head that could be interpreted as either a duck or a rabbit, but not both at the same time. Knowledge of this phenomenon goes back at least as far as Ancient India.

The figure below shows a relief carving on the wall of a Tamil temple.[5] It is 900 years old. What animals do you see in the image?

Fig. 3: Which animal do you see?

If you look at the image for a while, you should see both a bull and an elephant, with their heads merged together. The bull's horn becomes the elephant's trunk. But recognising this requires an expectation.

Not until we recognise the columnar legs of the creature on the right as those of an elephant, and the hindquarters and tail on the left as those of a bull, do we make out the two creatures' heads. But we can never perceive them both at the same time. This is because we do not recognise the heads by their shape, but rather the opposite; we read the shapes on the basis of an assumed specific interpretation—elephant or bull. For Wittgenstein, seeing always involves perceiving the world 'as something', or, in other words, perception always takes place on the basis of an expectation. On the one hand, the power of our expectations leads us to misinterpret facts. On the other hand, it often causes us to want to stick with an unsatisfactory state of affairs rather than accept change.

At best, this can sometimes help us correct a false expectation. This can only happen, however, if we recognise the expectation for what it is. Wittgenstein even coined a term for people who fail to realise that every point of view is built upon such preconceived expectations. He called them 'aspect blind'. Such people confuse their perceptions and thoughts with reality. But all of us, at least occasionally, suffer from 'aspect blindness'.

Second Illusion: We love novelty

I had my first-ever beer on a rainy school trip, and I will never forget how disgusting it was. We had an hour to wait for our train home and, rather than stay out on the freezing cold platform, we six boys went into the station restaurant. When the waitress appeared at our table, my best friend ordered with a confident voice: 'A lager.'. I didn't even really know what that was. As the youngest in my class, I hadn't even turned 14. The five pairs of eyes around the table turned to me. Did I even have a choice? 'The same,' I piped up bravely. The waitress didn't have a problem with our age: we were in Bavaria, after all. All she asked was: 'Halves?' Again, I didn't know what she meant, but my friend nodded. So I nodded, too.

    I had just enough time to glance at the neighbouring table and see the grownups clinking their beer glasses before the waitress returned with her tray. I sipped at the frothy head and earned some disapproving looks.

It tasted awful. It was bitter, sickly, and sticky all at the same time. I somehow managed to empty my glass with gritted teeth. I have no memory of the effect it had on me, and I don't know where our teacher was at the time. But I do know that it only took a few rounds of beer at the next few opportunities that presented themselves to persuade me that there is no better drink in the world than Augustiner Edelstoff lager. It was Bavaria, after all.

Why we love what we know

No one persuaded me to change my taste. No one needed to. I didn't even consciously change my preference myself. It just happened. I had a beer a couple of times, despite not liking it, and it wasn't long before the mere exposure effect kicked in. This is one of the most solidly researched effects in psychology. It describes the phenomenon by which people tend to develop a liking for things merely because they become familiar with them. Preferences arise through simple repetition.

We see this effect in all aspects of our lives, and it involves all our senses. It determines what we find beautiful or ugly. It gives rise to loving relationships and keeps them alive. Friendships owe their existence to this effect. Advertising, politicians' faces plastered on every lamppost, radio stations that play only pop hits, and the rituals of religious practice all make use of the mere exposure effect.

People often choose what they're familiar with, even when they are seeking a diversion. Feeling lonely at home in the evening? Go to your local restaurant and treat yourself to your favourite meal. Many people spend their vacation in the same place for decades: a place where they know every stone along the hiking trails. And have you ever asked yourself why the place you call home seems to be one of the best places on Earth?

The American psychologist Robert Zajonc is credited with discovering this effect.[1] Around 1960, he started researching the effects of the mere repetition of trivial stimuli. He showed young American test subjects Chinese characters that were meaningless to them. The subjects were exposed to some characters more often than others, and the additional familiarity alone was enough to give rise to preferences. The more often a certain symbol appeared, the more it was preferred. And that was not all: the subjects' mood improved when they encountered a more familiar character—in the same way that hearing a familiar hit song raises radio listeners' spirits.[2]

Zajonc's numerous scientific successors have replicated the effect with every conceivable stimulus. Repeated exposure to portraits of strangers made those depicted appear more friendly; oriental folk music became more appealing to Western ears with repeated listening; controlled experiments even measured how much better a dish or a drink tastes the more often it is served.³ It is not only humans who love what they're used to. The effect can even be seen in fertilised chicken eggs. After scientists had played various sounds to the eggs, the hatched chicks later showed a preference for those sounds.⁴

Fear is not the answer

Zajonc speculates that everything new is seen as a potential threat. The familiar appeals to us since, like every living thing, we require safety. If a given stimulus was previously encountered in a non-threatening situation, memories of that situation become associated with the stimulus, which in turn means we experience that stimulus as positive. Stimuli that are not associated with memories — that is, unfamiliar stimuli — are generally rejected. Does this mean that all organisms instinctively resist change?

Zajonc's theory sounds plausible, but it leaves a lot of questions unanswered.

Firstly, preference for the familiar is too ubiquitous for it to be explained purely as a fear of unknown dangers. Why should rockers prefer to listen to heavy-metal music over Mozart's *Eine Kleine Nachtmusik*, if their aim is to relax? And I don't know what fear response might cause a Bavarian native to yearn for a Munich-style beer in a Cologne pub.

Secondly, Zajonc himself was unable to provide a plausible explanation for his perhaps-surprising results. People prefer stimuli they have encountered before, even if they don't consciously recognise them. In some of his experiments, Zajonc flashed the Chinese characters for only a couple of thousandths of a second, so the subjects were only aware of them subliminally. The flashes were so short that the subjects were not even conscious of them. Nonetheless, their feelings about a character became more positive the more often they were exposed to it. The subjects did not know the source of their improved mood.[5] Furthermore, unnoticed rapid repetition led to a stronger improvement in mood than exposure to characters that subjects were consciously able to recognise, having been shown them for longer.[6] We prefer things we have already seen, heard or tasted, even when we don't consciously recognise them. This is one reason why calming memories are not enough to explain the mere exposure effect.

Homo explorans
Thirdly, Zajonc's theory overlooks the fact that, although the unknown can entail danger, it can also hold promise. Evolution has brought this to bear by balancing our fear of the unfamiliar with a powerful counterweight. Curiosity is one of our strongest drives as humans. It is such a vital trait because it allows us to learn.[7] Curiosity

not only enables us to handle the unfamiliar; it makes us actively seek it out.

If it weren't for their curiosity, our ancestors would never have crossed the oceans and settled new continents, mountaineers would never have scaled Mount Everest, and astronauts would never have set foot on the Moon. The genus *Homo* would have remained on the grasslands of East Africa as an insignificant branch of the primate family. Curiosity fuels change. We have curiosity to thank for everything that has made our civilisation what it is. It makes scientists build ever-bigger telescopes to wrest the secrets of the Big Bang from the night skies, or to seek to decode our own genome. It was curiosity that led Leonardo da Vinci to paint the *Mona Lisa*, and Ludwig van Beethoven to spend years perfecting a single symphony.

Not everyone has the urge to sail around the world, and not everyone is a born explorer. People vary greatly when it comes to how curious they are. Psychological tests show that openness to new experiences is one of the most stable personality traits. People who show curiosity in certain circumstances are more likely to be inquisitive in general. For traits such as conscientiousness or agreeableness, a person's reactions are far more dependent on the given situation. This is why it is often unfair to judge someone as generally pedantic or thoughtless—but people who are naturally more open to new experiences, or the opposite, certainly do exist.[8]

Of all personality traits, openness to new experience is the most inheritable. The more curious someone is, the more likely they are to have had curious parents. There are many indications that this variation is due to a gene that controls the construction of certain dopamine receptors in the brain.[9] Paradoxically, however, curious people are also subject to the mere exposure effect. Although they seek out new experiences, they also prefer the familiar. But when they are exposed to familiar stimuli too often, they become bored more quickly than less curious people.[10]

And what about those of a more anxious disposition? If the mere exposure effect could be explained as fear of the unknown, surely such people must be particularly drawn to the familiar. The opposite turns out to be true. Irrespective of whether they are generally more anxious or just feel anxious in the situation they find themselves in, fear suppresses the mere exposure effect and the preference it creates for the familiar.[11] This disproves the theory that we always avoid the new because it is scary.

'Mind at ease puts a smile on the face'

If it is not fear of the unknown that makes people prefer the familiar, what is it then? Scientists finally discovered the answer in experiments in which they varied the ease with which stimuli could be processed. In one such study, subjects had electrodes attached to their cheeks that could detect even the slightest twitch of a smile. They also had electrodes pasted near their eyebrows and the corner of their eyes to register the unconscious muscle movements that cause laughter lines. The test subjects were then briefly shown pictures of neutral everyday objects—a house, a car, a chair. In some of the pictures, the lines were imperceptibly emphasised, while in others they were slightly blurred. The clearer the lines were, the more likely the electrodes were to register a slight smile, and the more likely the subjects were to express a particular liking for the object depicted. They also preferred images that were displayed for a few tenths of a second longer than others. Clearly, the easier it was to perceive an image, the more the subjects liked it. Our brains reward us with positive feelings when we save it energy. This explains the title of the authors' paper: 'Mind at ease puts a smile on the face'.[12]

Other experiments demonstrated how this is connected to the mere exposure effect. If subjects were subliminally exposed to the outlines of a picture for a couple of hundredths of a second before being asked to judge the image consciously, they were not only likely to recognise it more quickly, but they were also more likely to find it attractive.[13] Also, when asked to predict the final word of a sentence, subjects found the answers more pleasing the more predictable they were. Thus, people can express a preference for the word 'yawn' over the word 'bead', if they are presented as the possible last words of sentences such as 'The bored student opened her mouth to ... yawn' [easily predictable] or 'The evening gown was missing a ... bead' [not easily predictable].[14]

Both those experiments go back to the principle of predictive coding described in the previous chapter. Our brains create an image of reality by constantly making predictions and then doing their best to find confirmation of those prognoses. If an already-familiar stimulus reappears, the brain's expectations are fulfilled. And if we see a bored student open her mouth, the prediction that she is going to yawn is a pretty sure bet. That the second sentence ends with the word 'bead', however, is not immediately obvious or predictable.

So this explains how our emotional responses can vary so much. Who, after all, doesn't like to be right about something? That pleasant feeling has far deeper roots than our simply liking to be right. If one of the brain's assumptions is confirmed, it does not need to form a new hypothesis—saving it work. This is the reason we prefer the familiar without knowing why.[15] People are curious—and, at the same time, they want to experience the world the way they know it. Thus, our consciousness is also split in this respect.

The joy of repetition

We always get out of bed right foot first or left foot first, but never the other way around. We stagger to the bathroom, eat the same breakfast as always, make our tea or coffee the same way as we do every day. Then it's back to the bathroom to clean our teeth, always starting with the same side of our mouth. We set off for work, taking the same route we have taken hundreds, maybe thousands, of times. We reach our destination without ever considering for a moment which way to turn at each intersection. We do it all on automatic. Only when a train fails to come or a road is blocked does our mind briefly return to our bodies from wherever it has been wandering so that we can make the necessary decisions.

Arriving at work, we head for the coffee machine as if on autopilot. With centimetre accuracy, we place our steaming cup of coffee in the same place as always on our desk before starting work. In the evening, our bodies find their way back to our front door, and we enjoy some leisure time until our hand places our toothbrush in our mouth in exactly the same way as ever, as if replaying a film from the morning. Then we climb into bed, with the same foot first as always.

'Ninety-nine hundredths or, possibly, nine hundred and ninety-nine thousandths of our activity is purely automatic and habitual,' wrote William James, the pioneer of both American philosophy and the scientific study of psychology.[16] Just as we find joy in a familiar view, and a favourite taste, or a much-listened-to piece of music, repeated habits also make us feel good. Family members always sit in the same place around the dinner table; football fans are overcome with joy every time they sing their club anthem before a game. Religion is not the only part of life that owes its appeal to our love of ritual.

People feel a sense of security when they know exactly what to expect. Repeatedly doing the same thing can also be a way to structure time. Habitual behaviours give us an impression of order, and can even help give meaning to our lives. This is the usual explanation, and it is by no means wrong.

But do we follow our habits purely out of a need for security and meaning? If that were so, there would be no reason not to vary our familiar habits sometimes. We could occasionally sit at the other end of the dinner table, or start brushing our teeth from the left-hand rather than the right-hand side. But we don't. People don't repeat their habits with minor variations. They always do exactly the same.

This is due to predictive coding: it saves work for the brain. Without routines, our brains would be hopelessly overwhelmed by the simplest of everyday tasks. This will be familiar to anyone who has ever learned to ride a bike, to dance salsa, or to cook a three-course meal. Our brain's ability to process signals is so sluggish that no novice can accomplish these tasks off the cuff. It takes hours of practice to cement the necessary programmes in the brain. Once that is the done, however, we are able to complete such tasks with an almost robotic degree of precision. And, since they are now so easy for us, we prefer such tasks to new, unfamiliar ones. Just as perception must be based on expectations, action is necessarily based on habits.

The human automaton

Habitual action has its own circuitry in the brain. While the cerebral cortex is responsible for making decisions in unfamiliar situations, centres much deeper in the brain handle routines. The basal ganglions in the middle of the brain control stereotypical actions.[17] They not only effectuate recurring patterns of movement, they also ensure that we stick to our habits.

Since habits are largely disconnected from the higher functions of the brain, they tend to persist, even long after they have lost any reason to remain. 'The pitcher goes so often to the well that it is broken at last.' The wisdom of this saying was proven by a series of elegant experiments carried out by neuropsychologists at the University of Cambridge in the nineteen-eighties. Working with rats, the researchers started out by training the rodents to push a lever regularly for a reward of some tasty grains. They then removed the lever and gave the rats the same food for nothing.

However, the researchers then taught the rats to hate the grains by lacing them with laxatives. Finally, the scientists returned the lever to the rats' cage. And what did the rats do? They pressed the lever over and over again, although they knew the grains would no longer give them pleasure, but a bellyache instead.[18]

Do humans behave any more wisely? What kibble is to lab rats, popcorn is to cinema-goers. In a field experiment carried out in the movie theatres of Los Angeles, test subjects were given popcorn that was either fresh or a week old. When the film began, almost all the subjects with fresh popcorn began snacking—but so did those who had been given stale popcorn (having been identified as habitual popcorn eaters when watching a movie). Those test subjects complained about the quality of the popcorn, but carried on eating it anyway.[19]

This is because the brain functions that control habitual action are fundamentally different from those directing deliberate actions. When we make a conscious decision to do something, it is in order to achieve a specific goal. A cinema-goer who only eats popcorn occasionally will reach into the box in the hope of either sating her hunger or experiencing pleasure. Her brain triggers the behaviour when there is a reasonable prospect of fulfilling that hope. But the processes at work inside the mind of the person sitting next to her, who regularly snacks on popcorn in front of the silver screen, are far simpler. Habitual behaviour is triggered by stimuli, not expectations. As soon as this man enters a cinema and smells the sweet aroma of freshly popped corn, he is certain to buy some. And as soon as he settles down in his seat, his hand will be in the box of popcorn—all without him even thinking about it. Just as the sound of a bell caused Pavlov's dogs to salivate, movies and popcorn are simply meant to go together for this man.

Our moviegoer might baulk at being compared to one of Pavlov's dogs. He might claim to buy popcorn because he likes the taste and the unique feel of it in his mouth. But the field experiment with the stale popcorn disproves that. The results of many other experiments also cast doubt on whether intentions play any role at all once a habit has become established.

What drives people to go to the gym? Data provided by gym owners gives the answer. It usually begins with a desire to do something for one's physical health and appearance. However, what actually happens does not depend on how strong someone's resolve is, but on how often the new members visit the gym in the first five weeks. Those who turn up regularly during that critical phase have a good chance of sticking with it in the long term. Thanks to the power of habitual behaviour, even those who continue to think of leg presses and triceps extensions as torture can come to consider their workout to be as much a part of their routine as brushing their teeth. Those who train less regularly in the first few weeks, perhaps due to work or other commitments, fail to develop a habit and will soon give up going to the gym.[20]

Zombies in the brain

We might argue with William James over whether 99 per cent, 99.9 per cent, or only 95 per cent of our daily activity is habitual. But no one disputes the fact that the majority of our actions are routine in nature. Thus, changing our lives means adopting new habits.

The good news is that it is easy; you just have to do it. Every regularly repeated action becomes a habit, as demanded by the principle of predictive coding. The less encouraging news is that the development of a new habit is almost always hindered by the existence of an old one. The difficulty is not creating routines; it is letting go of old ones.

So the first thing to do is to make space. However, that will not succeed as long as we continue to misunderstand what drives us. All too often, we mistake habits for goal-oriented actions. People who don't exercise believe they are unathletic, and prefer lazing around to sweating with effort. People who eat meat every day say it tastes good. Those who commute to work by car, even though they could make the journey just as quickly by train, claim that public transport is inconvenient. This makes behaving differently an unattractive prospect. Even though we know it would be better to do some exercise, eat more greens, or take the train, such behaviour is contrary to our preferences. Or so we think.

The pattern is the same in all three examples. First, we make the correct assumption that doing action A *can* be a way to achieve goal B. From this, we infer that we do action A *in order* to achieve goal B. But that conclusion is false. If I grab a hammer, I *can* use it to smash a window. But the likelihood is that I've fetched a hammer because I need to hang a picture.

This example is based on the assumption that I picked up the hammer with a particular intention in mind. When we act out of habit, however, we assume an intention that doesn't exist. Habits have a life of their own. Cinemagoers do not munch on popcorn because of the taste, and certainly not because they are hungry, but because they have developed the habit of snacking in front of the silver screen. Couch potatoes stay on the sofa because they are accustomed to spending their weekends without exercising. The Indian American neuroscientist V. S. Ramachandran described the countless routines we play out every day without wasting a thought on their purpose to alien creatures at work inside us. He called them 'zombies in the brain'.

It is understandable that we convince ourselves that our preferences are the explanation for the workings of the zombies. None of us likes to think our actions are controlled by demons, even if they are our own demons. But as long as we continue to persuade ourselves that the tyranny of old habits is actually our own will, we will not be able to escape them. The zombies lose their power when we see them for what they are.

Why campaigns and good resolutions fail

Routines respond to triggers, not to logical arguments. That is why good resolutions almost always fizzle out straightaway. It is also the reason why campaigns to convince the public to change their behaviours usually also fail. In the early nineteen-nineties, the US National Cancer Institute launched a nationwide campaign to persuade people to eat five portions of fruit and vegetables a day, and it gained a lot of attention. In a short space of time, there was a rise in the number of people who were convinced that it would be a good idea to eat an apple or a carrot every now and then. But it didn't go beyond the *would*. There was no significant rise in the consumption of fruit and vegetables.[21]

The zombies are roused by key stimuli. People who regularly eat popcorn at the cinema will start thinking about the buttery snack in the queue for tickets, if not long before. Those who eat meat every lunchtime can be totally convinced of the benefits of vegetarianism, but as soon as they smell meat grilling in the works canteen, they will be irresistibly drawn to the sausages and the chops. And that's the point at which the zombies take over.

There is a brief moment when such automatic responses can be halted. If the key stimulus is recognised quickly enough, it is still possible to stop the routine with a conscious decision.[22] This strategy has proven to be effective, and is even more likely to succeed if the critical moment can be filled with an alternative action.

Someone who wants to change their diet but finds bratwurst and schnitzel irresistible can start by visualising the situations in which they usually choose the meat option. This makes them aware of the behavioural trigger. It is then helpful, as a second step, to plan a new reaction to the smell of meat cooking — for example to head straight for the salad bar and then, as a reward, to the desserts counter.

This was the principle behind a programme to curb obesity among children and young adults first trialled in French cities.[23] While the American campaign merely spread information about fruit and vegetables, the trial organisers in France sought to change situations and routines. They held 'breakfast classes' for preschool and school children, in which students explored a balanced diet in a playful way. Using hip-hop music, the programme taught the children that exercise can feel good. Rather than having their food served up to them in the normal way, the children were allowed to choose their own meal from salad bars. These and similar methods led to a 50 per cent reduction in obesity among children and young adults in the participating cities. The programme has now been introduced in a number of other countries from Belgium to Australia.

How to change your habits

Automatic behaviour is also the reason we waste so much time on our phones. When a text message arrives just when boredom has set in, or when we just want to escape the unpleasant thoughts in our heads, we automatically reach for our phone, powerless to resist despite all our good resolutions. People in Germany spend on average around 40 hours a month on social media channels alone.[24] That's a full working week! (The global average is even higher, at roughly 70 hours per month.[25])

Scientists at the Max Planck Institute for Human Development in Berlin have discovered a simple yet highly effective remedy for this habit.[26] They used an app that briefly delays the opening of platforms such as WhatsApp, and TikTok. The user's screen goes dark, and a message appears instructing the user to breathe deeply for a moment and consider what the point of opening their phone is. The forced pause only lasts one second, but it is enough to break a user's routine. Initially, the effect of the brief pause for thought is to persuade people to put down their phones more often.

After six weeks of using the app, at the latest, users reach less often for their phones in general. A new habit has set in. Whenever users feel the urge to switch on their electronic time-robbers, they automatically ask themselves what the point of opening their phone is. The apps that chronically robbed phone owners of so much of their attention tend to be opened about half as often.

Users can even track their progress. They are shown how many times they resisted the temptation to access Instagram, YouTube, etc., over the previous 24 hours. They can also see whether they have been more or less steadfast than the previous week or month, or compare themselves to other users of the app. This kind of information is an additional help with behavioural change. Those who wish to change an old habit must leave their comfort zones. However, the small amount of effort it takes to stop the zombies is usually not rewarded until later, when we gradually realise how much more agreeable our new freedom is. Feedback about successes helps bridge that time lag.

Such positive feelings of success can work wonders when it comes to establishing new behaviours, as proven in an experiment involving employees at a hospital in America. Maintaining hand hygiene is a constant problem for clinics around the world, and the place where cleanliness is most important — the intensive care unit — is the ward where employees tend to wash their hands least often. Even placing clearly visible motion sensors and video cameras at every scrub sink does nothing to improve this. At the hospital on the east coast in which the study was carried out, doctors and nurses washed their hands before patient contacts less than 10 per cent of the time. That's a typical hand-hygiene rate.

The hospital's management then installed digital noticeboards in the hallways, showing how many times staff members had washed their hands during the current shift, with a comparison of their previous week's performance, and with that of other shift teams. Every time someone turned on the tap, the score went up. Handwashing rates suddenly changed, jumping from less than 10 per cent to more than 90 per cent.[27] And they remained at that high level. Washing their hands before any patient contact had become second nature to the staff. Their negligent behaviour had not been changed by warnings or admonitions, but by positive feedback, which reminded them constantly of the need to change their habits and encouraged every small step in the right direction.

The road to freedom

A decade and a half after my first beer, I moved from Bavaria, in the far south of Germany, to Hamburg at the other end of the country, where the traditional beer is pilsner. Topped by mountains of foam, this brew tasted stale, flat, and so bitter to me that I imagined an entire field of hops had been boiled down to make each glass. Of course, I was tempted to leave well enough alone and stick to Bavarian wheat beer, which was also available in Hamburg. On the other hand, I was curious to know why my friends loved this beer so much, and perhaps I wanted to feel like a Hamburg native myself. Now, after spending half my life in Germany's biggest North Sea port city, I have learned to love pilsner, and wouldn't drink Augustiner Edelstoff if you paid me. I long ago ceased to understand how people in Munich can call that sickly sugar-water beer.

New preferences and habits do not come from understanding alone. Breaking the tyranny of routines requires patience, as well as a strategy for reconditioning the brain. But understanding the problem is a necessary first step on the road to freedom. The question is not whether we can change ourselves. We can. The real question is, what makes us want to do it?

Third Illusion: It's always turned out fine before

One of the most beautiful and insightful passages from ancient philosophy tells of the importance of knowledge as a prerequisite for righteous behaviour. It is a dialogue between Meno, a distinguished general, and Socrates. Meno wants to know if correct behaviour can be taught.[1]

Then Socrates poses the key question: what is knowledge? Does knowledge mean knowing all the facts about a problem and having a correct opinion about those facts? Or is it more? Socrates asks further:

> 'If a person who knows the road to Larissa (or wherever else you like) should walk there and guide others, no doubt he would guide correctly and well?'
>
> 'Certainly.'

'And what if there's someone who opines correctly which road it is, but has neither gone there nor has knowledge—wouldn't he, too, guide correctly?'

'Certainly ... I wonder, Socrates, if that's the case, why in the world knowledge is so much more honoured than correct opinion, and why one of them is one thing and the other is another.'

'True opinions, for as much time as they stay with you, are a beautiful thing and accomplish all good things. Yet they're not willing to stay with you for a long time, but they run away out of the human being's soul; as a result, they're not worth much, until someone binds them ... But once they're bound, they first become instances of knowledge, then steadfast. And that's exactly why knowledge is a thing more honoured than correct opinion ...'

'By Zeus, Socrates, it is something like that!'

The last snow

My personal road to Larissa was in a high mountain valley in the Swiss canton of Grisons. It was February 2022, and we had taken a week-long trip to the Swiss Alps to do some skiing. We planned to spend our time climbing the high peaks with our climbing skins on and skiing back down them over the deep, powdery snow.

We were longing to see the winter scenery. There had not been a single flake of snow for months in Berlin, where I live. If there was anywhere in Europe that could guarantee snow-covered forests and biting cold air, it was here in the remote area beyond the Fuorn Pass, nearly 1,700 metres above sea level, where the weather stations still register 250 days of subzero temperatures per year. Or so we thought.

When we arrived late at night after a long journey, there was no sign of frost. Water was dripping from the roof of our accommodation: melting ice. The next morning, we were greeted by a clear blue sky. We looked up at the mountains. The forests were green; the meadows above the tree line were yellow, with grey rock rising above them. Here and there, little patches of snow clung on to the shaded sides of mountain gullies, many of them covered with mud from landslides. There was a better chance of witnessing winter scenes down in the valley. After breakfast, we had a coffee on the balcony in just our T-shirts, while anemones were already flowering in the meadows below us. The local people told us they had never seen a February like it. It had not snowed since before Christmas, and every day had been sunny since then. Like early summer, they said.

I felt somehow disturbed, in a way I had never experienced before. At first I put it down to a failure to contain my disappointment over our thwarted plans. But could a couple of cancelled ski tours really upset me so much that I couldn't sleep at night, and make me so absentminded during the day that I once even called one of my skiing friends by my wife's name?

After a couple of days, it dawned on me why I was so troubled. I was doubting my own eyes. My inability to believe what was happening before my eyes goes back to my childhood. I grew up in the Alps, and was familiar with the changing seasons in the mountains from a very early age. In winter, the snow glinted on the slopes; in late spring, it retreated to the higher peaks; and in summer it was only found on the glaciers. This was how it had been since long before we humans appeared, and this was how it would always be. Nature in the mountains was majestic and eternal. Green mountains in February were more than unthinkable for me. They were impossible. So what I was experiencing could not be real; it must have been a bad dream.

Even more disconcerting was the fact that my disbelief was in contradiction to what I knew to be true. I was familiar with the calculations of climatologists. I began taking their warnings seriously decades ago. One of the first articles I ever had published was about the danger of the Gulf Stream collapsing due to the effect of greenhouse gas emissions, which would pitch Europe into climate chaos. No one was talking about rising temperatures back then.

Climate change was real and disturbing, of that I was sure. And I was in little doubt that the rising temperatures, storms, floods, and droughts would probably spell disaster for billions of people.

So why was I so shocked to see the green mountains? Quite simply, I did not count myself among those billions of people. The adversities would be faced by others; I and my loved ones would be spared. I was sure of that, too. All we would face would be warmer, longer summers. And is it so bad to be able to go swimming in the lakes of Brandenburg in October?

On the sunny disposition of the Germans

To my credit, I was not the only person with such optimism. Germans are accused notoriously often of grumpiness. But, like all prejudices, this one is only partially true at best. It is true that most people in Germany take a dim view of their country's future. In a survey, 56 per cent agreed with the opinion that Germany was in decline. No fewer than 84 per cent were of the view that future generations would be worse off than themselves.[2] Those figures are alarming, but anyone who sees them as a reflection of a specific kind of German moroseness is mistaken. The corresponding figures are even higher in almost every other large European country.

However, when people are questioned about their personal views, there is little sign of impending doom. Sixty-three per cent of respondents in Germany even say they are optimistic about their own future.[3] The climate, the economy, migration, our ageing society—our consciousness is split when it comes to any of these problems. The general situation is somewhere between bad and devastating, but, despite this misery, we see our own future as rosy.

The overwhelming majority of Germans are also convinced that they are healthier, and, of course, that they are better drivers, than the majority of Germans. During the Covid-19 pandemic, most people assessed their own risk of infection to be lower than average. Couples know that half of all marriages fail, but happily tie the knot in the belief that they won't be facing each other in the divorce court within a few years. All these figures are well documented.[4]

Is this because we generally tend to overrate ourselves? It surely cannot be just that. We also live in hope that things over which we have no control will turn out best for us. For example, most people believe their risk of getting cancer is much lower than it really is. A rather curious study, but one that should be taken seriously, showed how Germans see the threats to their beloved beer gardens. Many see the damaging effect of the leaf miner moth, which attacks and kills horse chestnut trees, as a danger for the continued existence of such open-air establishments. However, despite all their concerns, they also believe the pest primarily attacks the trees in other parts of the country. Those questioned even judged the situation in their own locality to be better than in neighbouring areas. They claimed that a certain amount of damage from the moth was visible in other places in their region, but the horse chestnut trees in their beloved local beer garden were thriving as magnificently as ever.[5]

Optimists have it better

Optimism does not mean viewing the world through rose-tinted spectacles. Optimists are certainly not oblivious to the evils of the world. They can be split into two groups when it comes to the way they cope with that knowledge. The first kind of optimist believes the world can be changed for the better—even though the future remains uncertain. This the kind of optimism that gives us the courage to make changes. Without it there would be no progress. It gives people hope.

The other kind of optimist has no particular interest in change. Such optimists believe firmly in the existence of bad luck, but they are also convinced that it primarily strikes other people. In other words, they believe the laws of probability do not apply to them.

According to a 2013 study, at least 80 per cent of people are optimists of the second kind—not just in Germany, but across the world. Parents from Albania to Zimbabwe were convinced that their children were better-looking than other people's kids, homeowners dismissed their huge mortgage debts as unimportant since the value of their property would increase with time, and the majority were confident that their life would be better in the future.[6] The only group that failed to show any positive expectations at all were those suffering from clinical depression.

'One puzzle of optimism is thus that people maintain overly positive expectations despite a lifetime of experience with reality,' writes Tali Sharot, a London-based neuroscientist. She demonstrated that our cerebrum processes information selectively in order to remain hopeful in the face of reality. She found this effect in centres of the brain that process incorrect predictions. When a person receives bad news, those centres react less strongly to the bad news than to information confirming their positive expectations—as if the brain were refusing to acknowledge unwelcome news.[7] Excessive optimism thus follows the principle of predictive coding. We need positive expectations because they motivate us to act. As discussed in the previous chapter but one, our minds tend to defend their expectations—even when confronted with reality.

People can even be optimistic about their own optimism. Most believe they are less prone to sugarcoating the world than others.[8] Winston Churchill was one of the few people who recognised the irony of such self-deception, declaring, 'I am an optimist. It does not seem too much use being anything else.' It was probably no coincidence that the former British prime minister was able to see through his own illusions; he suffered from depression for much of his life.

In any case, modern research backs Churchill up by highlighting the benefits even of unrealistic optimism. Optimists, whether of the first or second kind, are not just happier; they are physically and mentally healthier than realists. They suffer less from stress, eat more healthily, and take more exercise. And, since believing that goals can be achieved motivates them to exert themselves to that end, they tend to be more socially successful and to be higher earners. People who overestimate their own prospects have a competitive advantage. The philosopher Daniel Dennet explains our tendency to see the world through rose-tinted spectacles as an evolutionary adaptation. For him, illusions such as those harboured by the second type of optimist are the only kind of false belief that were advantageous in our natural history, and they thus prevailed.[9]

Complacency breeds carelessness

What is good for an individual is often disastrous for a group. History can be seen as a series of disasters caused by overoptimism. The *Titanic* was supposed to be unsinkable, and in 1914 the German government and its generals were convinced that the war would be won within weeks and with few casualties. In 1938, the prime ministers of Britain and France believed signing the Munich Agreement would appease Hitler. After the Second World War, physicists promised us safe, almost limitless, nuclear power. In the first decade of this millennium, Wall Street traders appeared to have come up with a magic way for all Americans to be able to buy their own homes, in the form of mortgage-backed securities. Banking supervisory bodies believed the hype, until millions of bad loans led to the global financial crisis of 2008. The list of similar disasters in the recent past is almost endless.

There is some evidence to suggest that the more uncertain the prospects are, the more we humans tend to display excessive optimism. In uncertain situations especially, individuals who overestimate their own chances and who bluff have a competitive advantage. Their more cautious peers lose out. In this way, a tendency to go all out at critical times could have spread through the population.[10] This would be expected, according to evolutionary theory at least. Just think of the overconfidence with which Europeans joined a war in 1914 that turned into the worst bloodbath the world had ever seen to that point.

Pride comes before a fall

In Germany's Rhineland region there is an old saying that translates as something like 'Everything turns out fine in the end.' Such optimism may be helpful, or even essential, in private life, but a healthy dose of scepticism seems more appropriate in the wider world. This does not mean we have to view the situation the world is in as hopeless. But it is advisable to retain some doubt about what our ingrained optimism makes us believe: that things will sort themselves out, or at least that we, personally, will be spared any major change because fate favours us.

We are currently more prone to such illusions than ever before. Reckless overconfidence is a threat for three reasons.

First, the world is now changing so rapidly that it is becoming increasing difficult to predict the future. Even predictions for the coming year are almost impossible to make nowadays. Insecurity fosters excessive expectations, so any kind of innovation can easily lead to a vicious cycle of euphoria and disappointment. This is the reason why stock market bubbles regularly result from the introduction of a new technology such as the internet—only to be followed by a crash. And which of us today remembers fondly our naive enthusiasm over the advent of social media in 2005? What remains of the hope that the people of the world would become brothers and sisters on Facebook, Twitter, and the rest, and that democracy would flourish around the world because of it? And not forgetting the frenzy after the fall of the Berlin Wall. From today's perspective, the hope that eternal peace would reign in Europe seems like a fairy tale from the distant past.

Second, constant connectivity enables individuals' illusions to be strengthened in company. The more encouragement people receive for their hopes, the stronger those hopes become. This is what leads to the 'irrational exuberance' described by the then-Federal Reserve chairman, Alan Greenspan, in a famous speech about the dizzying heights reached by share prices during the internet bubble before the turn of the millennium.

Third, many of the issues that threaten us today are global. But all around the world, people believe that fate favours them over their local and global neighbours. This trait is particularly clear when it comes to ecological issues. Two large-scale research projects included questioning people from all continents about their personal assessment of climate change, air pollution, overpopulation, deforestation, and similar issues.[11] Respondents were asked to make comparisons. Was nature in a better or worse state in their own region than in the rest of their country? And was it better or worse in their country than elsewhere in the world?

From villages to megacities, from Mexico to India or Nigeria, from America to Australia, the answers were the same as those heard in the beer gardens of Germany. People everywhere believed that the environment in their own region was in a better state than in the rest of their country, and, compared to the devastation in other places, they believe their own country was doing okay.

Anyone who has inhaled the toxic air of an Indian metropolis, or experienced a deadly drought ion the African steppe triggered by inexorably rising temperatures, or seen images of the damage wrought by hurricanes in America might be surprised that the people in those places continue to extol the benefits of their home. This illusion is more than a harmless curiosity of human behaviour. The optimistic opinion that we are favoured by fate leads to societies shirking their responsibilities. When the foundations of our way of life are more damaged elsewhere than here, it is up to the people there to take action first. Because everyone feels this way, nobody does anything. After all, everything's always turned out fine — so far.

Death is always lurking somewhere else

Most people are more afraid of perishing in a plane crash than of being killed at the office by a ballpoint pen. However, in 2023, only 72 people were killed in air travel accidents worldwide. By contrast, the danger of choking on an accidentally swallowed pen lid is real. So real, in fact, that the International Organization for Standardization issued its norm ISO 11540. This stipulates that ballpoint pen caps should have a small hole at the end, to allow air through in the case of a choking emergency. However, not all manufacturers follow this regulation, leading to as many as 300 deaths per year in Germany alone due to choking on ballpoint pen components, according to the accident insurance company for public workers in the federal state of Lower Saxony.[12]

We hear and acknowledge such figures, but they remain abstract numbers for us. No one with even the slightest fear of flying would be persuaded to change their mind because of such statistics. We feel that the two risks are just not comparable. The reason for this is that we systematically underestimate dangers when we judge them on our own experience, and overestimate threats we hear about from others.[13] We encounter ballpoint pens every day without incident. Plane crashes are beyond the experience of most of us. However, we are all familiar with dramatic images of air traffic accidents in the news media. That gives us the impression that such disasters are more likely to happen than they actually are.

We make the same distinction between environmental degradation in our own locality compared to elsewhere. We become accustomed to the circumstances at home, so much so that a degraded natural environment is now the norm for us. Measured by biomass, three-quarters of insects have disappeared since 1990, even in protected areas in Germany, but the decline occurred gradually while nobody noticed.[14] As for the human toll: 28,000 people die in Germany every year as a result of particulate matter in the air, mainly from industrial chimneys and diesel motors; the heat wave of 2022, which was caused at least partly by climate change, claimed 4,500 victims alone; and for many years now, Germany has seen an unusually large number of deaths due to heat in comparison to other countries. All this is seen as normal, if regrettable.[15]

However, images of tornados tearing through American towns, melting icebergs in the Antarctic, and bushfires around Australia still leave us shocked. This further feeds the illusion that environmental disasters are on the rise everywhere except outside our own front door.

From facts to knowledge

Do you remember the dialogue between Socrates and the general Meno? The text was written by Socrates' student, Plato, more than 2,500 years ago, and analyses this issue with astonishing clarity. Socrates differentiates between knowledge and mere information. If you know the facts, you can form a true opinion. But opinions are 'fleeting', as Socrates puts it. We can follow them or not. Information is merely material for the mind to play with. It enables us to take the correct course of action, but it doesn't guarantee that we will.

People are unable to gain a real understanding of a

problem until another mental process has taken place. The information must be 'fixed' in the mind. Personal experience or intensive reflection consolidates facts into knowledge. And there is no getting around knowledge, since it encompasses more than just information; it arises out of experience. We have to travel the road to Larissa, the path to knowledge, ourselves. Those who make the long journey from Athens to the city at the foot of Mount Olympus will never doubt their memory of the landscapes of Thessaly, the smell of the fig trees, and the dust of the road.

Back in the green mountains of Grisons, I suddenly understood what Socrates meant. I had read hundreds of scientific articles about climate change. But I had always been able to console myself with a little dose of optimism—surely it wouldn't be so bad? Then I had no difficulty dispelling any fears of being personally impacted. I was familiar with the problem. But the problem didn't feel as if it were mine.

I don't know what shocked me more: the extremely unusual heat for February in the high Alps, or my own shock at it. Until that time I had been certain of knowing the full impact of climate change. But if I had really understood it, would the sight of those snowless peaks have alarmed me as much as it did?

I had more than enough information, but it had always been abstract. What I lacked was knowledge. Now here was knowledge, albeit painful knowledge.

For almost 30 years, I had written books, published newspaper articles, given talks, and taught at universities. All out of a conviction that science was making great progress that most people never even hear about, and that we would live in a better, fairer, and more sustainable world in future if more people could embrace the insights gained from scientific research. The key, I believed, was clear thinking and access to information.

Now I realised the limits of this doctrine of salvation. It didn't even work reliably for me — at least not when it came to climate change. For all my factual knowledge about melting polar icecaps and retreating glaciers, or about tipping points in the dynamics of the Gulf Stream and the insanity of producing more and more oil from shale rock, I had failed to notice that it had all been happening for some time already. Global warming was already shaking up the seasons. And the drama was no longer just playing out in the Antarctic, or the Caribbean, or on the edge of the Sahara Desert; it was happening in our temperate part of the world, too. The threat was right on our doorstep.

I had overlooked something that Socrates taught Meno in his 2,500-year-old dialogue: facts can inform us, but only knowledge can give us orientation. And personal experience is what turns facts into knowledge. Ironically, I had a similar experience with *Meno* itself. I had long been familiar with the dialogue, and had even given a university seminar presentation on it. But it was not until that experience in the high Alpine valley, when spring had already burst forth in mid-February, that I understood that the dialogue was also about my own work.

Gods at a loss

We are better informed than any human generation before us. Entire libraries are available to us at the push of a virtual button; online media report within seconds and in great detail about events half a world away; and social networks inform us constantly of the latest life events of those close to us and of people we have never met.

If our great-grandparents could travel through time and observe us in the present, we would seem omniscient to them, maybe even godlike. They would undoubtedly be envious of our screens. Gradually, though, the brighter ones among them would begin to realise that these developments are not just a blessing, but also a curse. The media take up our time and attention, but do not offer us experience. All we get is second-hand reports. And such a rehashed version of reality rarely leads to knowledge. Paradoxically, it is precisely this overabundance of information that leaves us increasingly at a loss.

However, we cannot simply do without all that data. In an ever more complex world, we need more and more abstract information in order to find orientation. The exponential increase in infections during the Covid-19 pandemic, the creeping warming of our atmosphere, the effect of algorithm-controlled news streams on our way of thinking—our senses register all this indirectly at best, and our intuition is not designed to take it all in. We need data, symbols, and explanations in order to understand events. This is the only way we can learn new facts. However, as long as our minds do not connect the abstract level with concrete experience, those facts will remain alien to us and therefore often unable to influence our behaviour.

In our age, more than ever before, we have a need to turn facts into knowledge. In our increasingly complex world, it is not easy to underpin information with experience. It can, however, be helpful to remind ourselves regularly of the fact that the opinion of even the cleverest of commentators can at best only prompt us to do our own thinking; it can never replace our own thinking. It is also helpful to keep in mind that information from a screen cannot replace seeing things at first hand. One single personal conversation with an immigrant can be more enlightening than watching hundreds of hours of TV discussion shows about migration, and a glance at the leaves on the trees in your local park is more convincing than any climatologist's charts.

Fourth Illusion: Knowledge is power

Advances in medical science are one of humanity's greatest achievements. The plague has been beaten, smallpox has been eradicated, and diseased hearts can be replaced. Women no longer have to fear for their lives when they give birth, as long as there are medical practitioners available.

We have collectively forgotten the risks entailed in childbirth just a few generations ago. Ignaz Semmelweis did remember. On 1 July 1864, which happened to be his 28th birthday, he took up his first position as a junior doctor at Vienna's General Hospital. He had chosen gynaecology as his specialisation because he wanted to provide free care to destitute women giving birth. However, the young physician's happiness at helping these women, many of whom were homeless or working as prostitutes, lasted just a few days. Semmelweis soon found that it was not the happy miracle of new life that awaited him on his ward at the hospital's First

Obstetrical Clinic, but death. The ward was full of young women who were dying. And the doctors could only look on in consternation as one after another met the same fate. A few days after giving birth without complications, the new mothers developed a high fever and pain so severe that they could only writhe in bed screaming until they finally died.

The work 'provoked in me one of those unhappy dispositions that make life unenviable', Semmelweis was later to report. 'Everything was in question. Nothing was explained. Everything was doubtful; only the great number of deaths was undoubtable reality.'

The events at the First Obstetrical Clinic were common knowledge. Pregnant women begged not to be sent there. Some risked everything by giving birth on the streets rather than go there. The ward where Semmelweis worked was said to be cursed. That was certainly unfair, but most women at the time were unaware that similar horrific scenes were playing out in many European and North American maternity hospitals. From Philadelphia to Moscow, thousands of new mothers were dying of puerperal fever (also known as 'childbed fever') every year.

Women who gave birth without medical assistance, either at home or on the streets, were at risk from complications — bleeding, umbilical cord prolapse, life-threatening seizures — but they almost never contracted puerperal fever. Another mystery was the fact that the Second Obstetrical Clinic in Vienna saw far fewer deaths than the First. Semmelweis examined the patients thoroughly himself. The death rate in his clinic continued to rise. He eventually began to wonder if the sight of a priest constantly giving the last rites to dying patients might be so upsetting for the new mothers that they developed a fever from the shock. He ordered the clergyman to enter the ward discreetly, without ringing his bell. Still, there was no change to the death rate.

Unusually, the fever claimed the life of a man in 1847. He was a forensic pathologist and a friend of Ignaz Semmelweis. His grief led Semmelweis to search for the cause of his friend's death. A student had accidentally cut him with a scalpel during a postmortem examination. A few weeks later, Semmelweis's friend was writhing in agony just like the women on the ward. Could it be that 'cadaveric poison' was responsible for the death of both the pathologist and the new mothers? The idea must have seemed absurd at the time.

At the First Obstetrical Clinic, autopsies of the women who died of puerperal fever were a daily occurrence. After such a postmortem examination, doctors and students would rinse their hands with a little soap—if they washed them at all—before starting their rounds on the postpartum ward, with 'cadaveric material sticking to their hands', as Semmelweis put it. Could they be transmitting the poison from the cadavers to the healthy mothers? That would explain why puerperal fever was raging through his ward but was barely present among the women who gave birth at home, on the street, or at the Second Obstetrical Clinic, where no autopsies were carried out. If that were the case, the doctors were inadvertently killing women with their examinations—most notably, Semmelweis himself. It was a horrifying thought. But Semmelweis was determined to explore every possibility.

He instructed his staff to disinfect their hands and all their instruments with a solution of chlorinated lime after every dissection. (He assumed the solution would destroy the cadaveric poison; Semmelweis did not know that infections are actually caused by bacteria.) And, indeed, after just a few weeks, almost no women were dying of puerperal fever. When a dozen patients contracted the disease nonetheless, Semmelweis realised that it could be spread by living people as well as cadavers. His doctors were then instructed to wash their hands with the chlorinated lime solution after every examination. And puerperal fever disappeared almost completely from his ward.

This gave Semmelweis hope for the first time. He wrote to the directors of maternity hospitals throughout Europe to inform them of his discovery. He explained, argued, and compiled meticulous figures that are still considered exemplary for medical research today. *The Journal of the Imperial and Royal Society of*

Physicians in Vienna reported his findings; the article called Semmelweis's work as significant a breakthrough as the invention of the smallpox vaccine half a century earlier.[1] With simple hygiene, puerperal fever appeared to be defeated.

The regulations that had saved the lives of so many mothers in Vienna could now have been introduced overnight in maternity hospitals throughout Europe. Semmelweis would have been celebrated as a benefactor of humanity to whom tens of thousands of women a year owed their lives. But nothing of the sort happened. Only two clinics—one in Heidelberg and one in Kiel—began requiring their doctors to disinfect their hands and their instruments. Every other hospital refused. They stuck by their practices as if Semmelweis had never existed.

The hospital chief physicians appealed to the leading medical authorities of the day. And those authorities, including the famous physician Rudolf Virchow, rejected Semmelweis's proposals. Puerperal fever had many causes, they declared unanimously. And physicians causing the death of their patients was certainly not one of them. Others pointed out that the young doctor from Vienna did not warrant any attention because his hypothesis was not new. Hadn't a Scottish obstetrician by the name of Alexander Gordon made similar claims 50 before? 'It is a disagreeable declaration for me to mention, that I myself was the means of carrying the infection to a great number of women,' Gordon wrote in a treatise on puerperal fever.

The Academy of Medicine in Paris rejected Semmelweis's ideas twice. The president of the oldest medical association in America, himself an obstetrician, declared, 'Doctors are gentlemen, and gentlemen's hands are clean.'[2] And so new mothers continued to die.

When pride kills

All doctors feel duty bound to help their patients to the best of their knowledge and ability. This has been part of the Hippocratic Oath for more than 2,000 years. Semmelweis's data were widely known, and the recommendation for doctors to disinfect their hands before examining patients was both urgently needed and easy to follow. How could an entire generation of doctors resist it? And how could the great medical figures of the 19th century—highly educated scientists who had benefited the study of medicine—overlook something so obvious?

Hundreds of thousands of women and, ultimately, Semmelweis himself were victims of the denial of reality. In the chapter on the first illusion, we saw how the human brain interprets facts in the light of our expectations. This is the principle of predictive coding. Our brains are continuously making predictions, around which they construct an idea of reality. If the prediction does not match reality, that idea should change. This is how we learn. But we often cling to our ideas of reality, even when they clash with the facts. Dealing with contradictions requires work, which our brains like to avoid. The desire to have our illusory assumptions confirmed is referred to as 'confirmation bias'.

The mechanism responsible for confirmation bias is the same as that which gave us the optical illusions in the chapter about the first illusion—the human brain stubbornly clings to its preconceptions. However, there is one difference: optical illusions occur completely unconsciously. Although the information that we could use to create our perception of reality does reach the brain, it is suppressed before an idea can even form in our mind. That's why it is impossible to see the middle square of the chessboard on page 48 as being as light as it actually is. Even when we know that we are succumbing to an illusion, our brains are unable to correct the mistake.

The confirmation bias that cost so many women their lives, on the other hand, occurred at a higher level of the mind. Anyone who wished to do so could completely understand why Semmelweis implored his colleagues to disinfect their hands before examining postpartum mothers. The doctors had all the necessary information, but they disregarded it. Unlike the UFO believers in the Brotherhood of the Seven Rays, their minds were not clouded by collective hysteria or mortal fear. On the contrary, these medics, on whose decisions the wellbeing of their patients depended every day, could be expected to have far superior judgement.

The tendency towards optimism discussed in the previous chapter is also based on confirmation bias. We refuse to acknowledge news that threatens our rose-tinted view of the world, even when all the facts indicating that fate is not going to be particularly kind to us are available. We simply dismiss them as unimportant.

A filter bubble in the brain

Neither knowledge nor intelligence offer protection against confirmation bias. On the contrary, 'the greater your cognitive capacity, the greater your ability to rationalise and interpret information at will, and to creatively twist data to fit your opinions', writes neuropsychologist Tali Sharot. 'Ironically, then, people may use their intelligence not to draw more accurate conclusions but to find fault in data they are unhappy with.'[3]

So, ironically, it was the doctors' mental agility that made them more prone to sticking with their convictions. They used their reasoning abilities to perform some mental gymnastics. Thus, it was no coincidence that those who discredited Semmelweis were the great medical experts of the time.

The doctors also used their knowledge to pick fault

with one detail of Semmelweis's theory. They pointed out that the young Viennese doctor's idea that some kind of 'cadaveric poison' was the cause of puerperal fever was imprecise and, as would later turn out to be the case, wrong. The disease is caused by the transmission of a bacterial pathogen, not poison. Criticism of this kind did not touch on the crucial point — that doctors were at fault for their lack of hygiene. But the objection was enough to undermine Semmelweis's credibility.

For one thing, the doctors did not question the facts themselves, but rather their significance. Finding reasons to believe that Semmelweis's data were correct but irrelevant was a way for the doctors to avoid questioning their own convictions or changing their behaviour. In his *Lehrbuch der Geburtshülfe* (*Textbook of Obstetrics*), the man who succeeded Semmelweis in Vienna claimed that puerperal fever had more than 30 causes; he placed infection second-to-last on the list.

Such reactions are typical of confirmation bias. Countless psychological experiments have shown that people almost always measure the value of a piece of information according to whether it supports their own beliefs or not.[4] Newspaper readers are more likely to find headlines relevant, and to engage more closely with the corresponding articles, the more they confirm the readers' own worldview.[5] This phenomenon, now known as a filter bubble, existed long before social media began serving up a biased selection of news to their users. The business models followed by digital corporations did not create it, but they exacerbated it.

Ultimately, the algorithms that control our newsfeeds are a digital copy of the filter bubbles in our heads. In a confrontation, we automatically take other people's arguments less seriously when they are opposed to our own opinions. The resulting activity can be measured in the brain.[6] When we encounter a contradicting opinion, the activity of certain centres in the prefrontal cortex of the brain is reduced—almost as if the brain were switching itself off. This effect can be observed even in the most inconsequential of conversations.

'The stiffer the coat the prouder the busy surgeon'

Semmelweis's data were far from inconsequential for his fellow doctors. They meant that medics had to change their routine. An even greater problem was the moral discomfort they caused. Among doctors of the 19th century, there was a widespread opinion that 'pus was as inseparable from surgery as blood', as the British medical historian and anaesthetist Richard Gordon put it.[7] 'Surgeons operated in blood-stiffened frock coats—the stiffer the coat the prouder the busy surgeon.' Gordon quotes Sir Frederick Treves, a famous London surgeon of his time: 'There was no object in being clean. Indeed, cleanliness was out of place. It was considered to be finicking and affected. An executioner might as well manicure his nails before chopping off a head.'

Semmelweis had shown that an unhygienic doctor was not a hero, but a danger to the life of his patients. Those who took the statistics seriously not had only their confidence as doctors shattered, but also possibly their reputation. This gave Semmelweis's fellow doctors all the more reason to cling to their convictions, even if they were wrong. Changing their minds would cost them too much.

The American social psychologist Anthony Greenwald spoke of the 'totalitarian ego' as an entity that controls our feelings and behaviour much like a dictator controls a nation.[8] Both want stability, and suppress any impulse or thought that might call their own infallibility into question in order to achieve it.

Doctors who followed Semmelweis's instructions had to deal not only with self-doubt and suspicion from others, but also with a very guilty conscience. The Kiel-based obstetrician Gustave Adolph Michaelis was one of the few adherents to Semmelweis's theory. He committed suicide when he realised how many women's deaths he had been responsible for.

Another aspect of the resistance to Semmelweis's ideas was typical of confirmation bias in another way. The more consequential a theory is, the more likely people are to filter the facts. They ignore not only the information that contradicts their opinion, but also that which might have unpleasant consequences. We deny problems exist when we don't like the solutions to them.

Solution aversion

Every political talk show on TV offers a platform for solution aversion. A politically right-wing guest might complain that the left refuses to recognise the problems resulting from immigration because left-wingers reject curbing migration as a solution. Conversely, environmental activists accuse conservatives of downplaying the dangers of climate change in order to protect industry.

People will assess the irrelevance or importance of the same problem differently depending on the prospective solutions to it. The speed at which such assessments can change was demonstrated in a series of studies in which Americans were asked to give their opinions on environmental issues and gun laws.[9]

In the first study, subjects were interviewed about their political leanings towards either the Republican or Democratic parties, before being divided randomly into two groups. One group then listened to a speech in which the speaker argued that free markets and unbridled capitalism could solve the climate problem. The second group heard the same speaker saying that only laws and controls could stop the global rise in temperature. All subjects were then asked to assess how realistic they judged a prognosis from the United Nations Intergovernmental Panel on Climate Change to be, which stated that burning oil, coal, and gas would heat up the atmosphere by an average of two degrees.

The majority of the Democrats agreed with the prediction. The Republicans who had listened to the speech about free markets solving the problem were also willing to believe the science. By contrast, however, the Republicans who happened to hear the speech about the need for regulations disputed the prognosis. They claimed that climatologists were exaggerating the facts, and asserted that the rise in temperature would surely be far smaller. Therefore, science scepticism cannot necessarily be explained by the fact that people have a fundamental mistrust of science. They just don't want to hear about uncomfortable research results.

Another study focused on a different question: did the participants believe the data published by American respiratory physicians showing that 44 million Americans are exposed to health risks due to particulate matter and other pollutants in the air? The results were similar: test subjects who trusted in the free market believed the physicians when the speaker they heard claimed that the invisible hand of market forces would clean the air. When the same committed capitalists were exposed to a speech threatening regulation, on the other hand, they disputed the data.

A third study approached the issue from the opposite political perspective, and dealt with gun ownership. Subjects were asked their opinion about the right of every American to own a gun. They were then asked to read a text about violence suffered by innocent homeowners at the hands of burglars. As before, the subjects were then divided randomly into two groups. The first group heard a speech arguing in favour of increased gun control: they were told that the nation suffered so much violence because criminals can get hold of firearms so easily. The speech heard by the second group argued for the right to self-defence: subjects were told that free access to guns had to continue because only a gun in every home could protect citizens from all-pervasive crime.

How did the subjects assess the danger posed by burglaries? Hardly surprisingly, the gun proponents believed it to be considerable. The gun-control advocates who had randomly ended up in the first group agreed. They had just been told that the root of the problem was the excessive number of guns in circulation. This fitted with their worldview. However, the gun opponents who were randomly assigned to the group which heard that the solution to the problem of violent burglaries was yet more gun ownership were of a completely different opinion, declaring that violence during burglaries was not a serious problem.

These results put paid to the hope that enlightened people behave reasonably and that they just need to be put in a position in which they understand the problem. Knowledge of the facts is not enough to change people or their behaviour. Indeed, we regularly treat any facts that might shatter our preconceptions the way that Semmelweis's fellow doctors treated his findings: we simply ignore them. Knowledge can be shockingly powerless.

Faster than the snail

Two years after going public with his findings, Semmelweis was removed from his position in Vienna. He moved to Budapest, and in 1861 published a 500-page book in an attempt to rouse is fellow doctors to action. Critics tore his book to pieces.

How could he have promoted his cause more successfully? Semmelweis was not a strategist. He failed to secure the support of influential Viennese doctors, who were quite sympathetic to his ideas at first.[10] He left it to others to publish his findings in academic journals, and the book he eventually published himself was not exactly an easy read. In these ways, he made it unnecessarily difficult for his fellow doctors to follow his lead.

Semmelweis reacted to the wave of open rejection he faced by going on the attack. He wrote a bitter open letter to the leading obstetricians of the day. In it, he promised to be a 'tireless opponent' of anyone who dared to 'spread dangerous fallacies about puerperal fever', referring specifically to the addressees of each letter. 'You, Professor, are complicit in this massacre. The killing must stop, and I shall stand guard to ensure the murdering ends,' he wrote to the Viennese gynaecologist Joseph Späth. Semmelweis's tone was reminiscent of the embitterment and bewildered despair with which direct-action environmental campaigners today seek to jolt their fellow human beings out of their state of lethargy—leading to deeper entrenchment on both sides.

Nonetheless, it remains doubtful that Semmelweis would have achieved the breakthrough he sought if he had been a better communicator. The idea that childbed fever was spread by doctors was not Semmelweis's only innovation; the way he reached his conclusions was also completely new. No medics before him had ever based their theories on statistics. The figures that Semmelweis used were more than convincing, but his colleagues were simply not accustomed to thinking in terms of frequency rates.

Semmelweis never recovered from the rejection of his book. He started drinking, and suffered from severe depression. In the summer of 1865, he was admitted to a so-called lunatic asylum near Vienna. Shortly thereafter, on 13 August of the same year, Ignaz Semmelweis died in unexplained circumstances at the age of 47.

It was at this point that the tide began to turn. The impetus for the shift came from an unexpected source. On 12 August, one day before Semmelweis died, a young Glasgow surgeon called Joseph Lister used antiseptic while operating for the first time. While performing a complicated operation on a child, Lister dressed the wound with a bandage drenched in phenol. The boy recovered remarkably quickly. Lister began to use the method routinely, and soon began spraying his scalpels, and even the entire bodies of his patients, with phenol. He observed the same effect that Semmelweis had seen on the maternity ward: disinfection reduced mortality rates dramatically.

Lister published his findings immediately. Still, he was given a similar reception to Semmelweis two decades earlier. He was attacked by his fellow surgeons, who even accused him of plagiarism. He was ridiculed at medical congresses, and even as late as 1873, the leading medical journal *The Lancet* warned its readers about his ideas. Even the nursing staff in his own hospital reacted with hostility to him and his new methods.[11]

However, Lister had two advantages over Semmelweis. First, he was able to explain convincingly not only that his measures worked, but also why. He had studied Louis Pasteur's experiments on fermentation and putrefaction, and learned from them. They showed that microorganisms were responsible for decomposition processes, and could also cause infections. Chemicals such as phenol killed those microbes. Lister leapt at Pasteur's famous experiments as they strengthened his position. Semmelweis, on the other hand, knew nothing of the germ theory of disease.

Second, and more importantly, Lister possessed a crucial character trait that Semmelweis had lacked—patience. He spent many years politely but firmly fending off one attack after another. This helped antiseptics finally to take hold in medicine. Gradually, the new hygiene practices were adopted by more and more young doctors who had not yet internalised the tradition that a soiled coat was something to be proud of. Visitors travelled from abroad to learn from Lister. When Louis Pasteur demonstrated in his Paris laboratory in 1878 that puerperal fever was caused by bacteria, gynaecologists finally dropped their opposition. This put an end at last to the high mortality rates in maternity hospitals, four decades after Semmelweis's discovery.

We do not know how Joseph Lister bore the years of knowing that his fellow doctors who were hostile, or even just indifferent, towards him were risking the lives of countless people. Maybe he took comfort in the knowledge that he had set the ball of progress rolling, even though he could do nothing to make it roll faster.

In his 1972 book *From the Diary of a Snail*, the German Nobel Prize–winning author Günter Grass describes an exchange with a child:

> 'What do you mean by the snail?'
> 'The snail is progress.'
> 'What's progress?'
> 'Being a little quicker than the snail …'

Fifth Illusion: Freedom is the answer to everything

The American mathematician John Nash was the first to understand the nature of situations in which no one can act without putting themselves at a disadvantage even when the situation is unsatisfactory for all concerned. This was in 1950. Such a situation is known as a 'Nash equilibrium'. It's a common occurrence today, which could be described by a newly coined adjective: 'Nash'. The nasty tone of social media exchanges is 'Nash', and holiday jets full of otherwise-climate-conscious holidaymakers are 'Nash'. Shopaholics who run up massive debts in their race to keep up with the Joneses are 'Nash', as are countries who threaten each other with nuclear weapons.

John Nash was just 22 years old when he came up with his insightful concept. He was a doctoral student in mathematics at Princeton University in the United

States. His theory was largely ignored at the time, and even Nash himself considered it insignificant beside the work he had already published on pure mathematics. Some years afterwards, he became ill with paranoid schizophrenia. He started hearing voices, seeing secret messages encoded in mathematical formulae, and believing that every man wearing a red tie was an agent of a worldwide communist conspiracy. He was admitted to a psychiatric hospital, where the doctors repeatedly placed him in an induced coma with massive doses of insulin. John Nash never published another academic paper, and many in the mathematics community assumed he was dead.

In reality, he was still battling the voices in his head. Over the course of three decades, he managed to conquer his paranoia. (The story of Nash's struggle later became the inspiration for the 2001 Hollywood movie *A Beautiful Mind*.) In the intervening years, social scientists recognised that Nash had provided them with one of the most fundamental insights into the behaviour of humans and other animals. John Nash had explained why societies often find it so difficult to introduce change, even when there is general agreement that change is necessary. He was awarded the Nobel Prize in Economics in 1994. Tragically, Nash and his wife died in a traffic accident in 2015, when the taxi they were in veered off the road as they were travelling back from the airport after returning from Norway, where Nash had just received the prestigious Abel Prize for mathematics from the Norwegian king.

Stuck in an equilibrium of horror

The daily battle on our roads is also 'Nash'. For example, commuters at rush hour would almost certainly get home faster if they all took the bus instead of blocking the roads with their cars. If they did, it would then make sense to provide bus services that could get anyone to their destinations quickly, conveniently, and reliably—and at a fraction of the cost of driving.

However, if only a few individuals switched to a bus, their journey would take even longer, since the bus would be stuck in traffic along with all the cars. In that case, it would be only reasonable for commuters to switch back to driving.

And even if everyone switched to a bus en masse, the now-empty roads would immediately tempt some to fetch the car out of the garage again. That is an understandable reaction, too. After all, most people feel more than overworked in their job, and want to dedicate their precious free time to family and friends, so they want to waste as little of it as possible travelling to and from work. Others would follow their example and also return to driving, and once again the roads would be hopelessly congested. (It is understandable that frustrated commuters demand wider roads, and more of them. However, experience has repeatedly shown that such a solution alleviates traffic chaos for a short time at best, since they eventually also attract more traffic.)

Just as a marble will always roll to the lowest point in a bowl, human interaction, if left to its own devices, will strive to reach equilibrium—and all too often, that Nash equilibrium turns out to be an equilibrium of horror.[1]

This is a classic dilemma in game theory. Developed in the first half of the 20th century, game theory is a branch of mathematics that studies the role of reasonable decision-making in situations of social conflict. Its name comes from the fact that it analyses behaviours as if they were moves and countermoves in a game. John Nash's work put game theory on a completely new footing.

The most famous thought experiment in game theory involves two gangsters who are being held for questioning in two separate cells. Prosecutors suspect them of having robbed a bank together, but have no proof. So they offer both suspects the chance of acquittal, if they turn crown witness in the trial of the other, who would then be sent to prison for ten years. If both squeal, they each go to prison for five years. If both keep mum, they will only face a six-month sentence for illegal possession of a firearm.

Clearly, it is better for each prisoner if the other

does not grass. And each of them keeping shtum themselves is the way to bring about the best result for both. If both refuse to talk, they both get away with a short prison sentence.

But is it advisable to do that? Even if one suspect rejects the prosecutor's offer in the interests of the other, he cannot be sure his partner in crime will do the same. Held in separate cells as they are, there is no way for them to coordinate their responses. The temptation to rat on the other and gain immediate freedom is great. And going down for ten years for being too trusting would be a bitter fate. So the only remaining reasonable option is to put the blame on the other suspect. Since he's in the same situation, he will also do the same. And both will receive five-year sentences. The outcome that would have been best for both—a six-month prison sentence—becomes unattainable when each factors in the possibility of being denounced by the other. In the prisoner's dilemma experiment, betrayal is 'Nash'.

The logic of the brawl

In the previous chapters, I looked at four illusions that cause people to misjudge the situation they are in despite having ample information, and thereby lead them to avoid change. Those four illusions result from the way our brains use our perceptions and habits to maximise efficiency. In this chapter and the next, I examine decision-making, and explore the reasons why humans adhere to situations that are not beneficial to them, even when they recognise them as being detrimental. This often goes back to the fact that we evaluate wins and losses differently—a phenomenon examined in the next chapter. But before this, we need to take a look at the way that competition stands in the way of change.

Once the prisoner's dilemma is clear in our minds, we start to recognise the same structure in many of the problems of human coexistence. Conflicts in our relationships follow this game theory as much as the causes of global environmental crises do. Fishing fleets, for example, plunder the seas until fish stocks collapse, despite the fact that every fleet operator has an interest in allowing stocks to recover for the sake of future fishing. However, any fleet that holds back simply opens the way for a more unscrupulous competitor to plunder the fishing grounds.

Many of the upheavals caused by the rise of computers and the internet follow the same logic. Everyone who uses social media, and society as a whole, would be better off if Twitter—now renamed 'X'—Facebook, and the rest would stop incessantly spreading disinformation and hate speech. But it is the users themselves who put those messages online. So why is it impossible to turn social media platforms into places of friendly debate, sources of reliable information, and instruments of harmonious coexistence? It's because anyone who tries to do so is hopelessly overwhelmed by more provocative content. Bluntly worded posts are the way to reach a substantial audience, even if that causes the internet to spiral into an even more extreme place, and even if it is eventually to the poster's own detriment in the long run. Ranting on the internet is 'Nash'.[2]

In such situations, the only rational decision is the one that has the worst outcome for everyone. We don't have to evoke the concept of sin, as the Apostle Paul did, to explain why people do things that they themselves consider reprehensible.

Sophia's choice

The logic of the prisoner's dilemma is the deeper reason behind the seeming madness of cognitive dissonance. In 2019, before the flooding disaster on the River Ahr in Germany, the drought emergency in Italy, and the devastating forest fires in Greece, 30,000 respondents, selected from every member country of the European Union to be representative of the general population, were asked whether climate change was having an impact on their lives. Eighty-two per cent answered 'yes'. An even bigger percentage said they believed their own children would feel the consequences of global warming in their future everyday lives. No fewer than 90 per cent of respondents agreed with that statement—49 per cent said they were certain of it, while another 41 per cent responded that they thought it was probable. There are few other issues about which Europeans are so united.[3]

However, the overwhelming majority of those interviewed also said that this was no reason to despair. They believed the situation to be very much in their hands. Sixty-nine per cent of Europeans said they thought their own individual behaviour could help tackle climate change, either 'to some extent' or 'to a great extent'. And tackling climate change was seen as urgently necessary, as it was considered to be one of the greatest challenges facing humanity this century.

Those figures are astounding, not least because we often hear the opposite message, not just in everyday conversations, but from the scientific community as well: the battle against the greenhouse effect is hopeless because any individual's impact is negligibly small, and because the polluters are in competition with each other. There is no incentive to change anything. The status quo is 'Nash'. Are those sceptics right?

In practice, the argument goes something like this: 'I know flying is a major contributor to global warming. But we deserve a holiday. And the planes will take to the skies whether I'm on them or not, because everyone wants to fly. It makes no difference to the climate if the plane that pollutes the atmosphere is carrying my family or someone else's. So why should we be the ones to give up our spring weekend in Mallorca?'

We are all familiar with this kind of reasoning. We would like to do the right thing, but the situation can't be helped anyway. Psychologists call this a lack of self-efficacy. When people doubt that their behaviour can affect a situation, they understandably don't even try. Of course, for a hypothetical holidaymaker — let's call her Sophia — it feels bad to be acting contrary to her moral convictions. But it feels even worse to make a sacrifice she cannot be certain will have any effect. So Sophia prefers to put up with the cognitive dissonance and enjoy a couple of pleasant days in the Mediterranean.

The simplest option for Sophia would be to sugarcoat the issue for herself. We have already seen how this works: full of optimism, Sophia could decide that climate change does not affect her. That would at least mean she needn't fear doing something that might be to her own detriment. She could also give credence to the claims of the airline to offer particularly climate-friendly flights, or she could just join the ranks of the climate sceptics from the outset. She could maintain her belief in her own integrity in any of these ways. All she would need to do would be to dismiss as irrelevant any information that undermined her view of the situation.

The wealthier a person is, the more they consume, thereby contributing more to climate change. Sophia also knows that the richest 10 per cent of Europeans are responsible for almost one-third of the continent's carbon emissions — the same amount as the poorest 50 per cent combined. The more someone earns, the greater their ability to help mitigate climate change. Sophia earns well. She is aware that she bears a special responsibility.

Questioning her own decision, she considers the pros and cons. The holiday break promises to deliver a benefit, but also creates costs. Sophia is not so naive as to believe that the costs do not go beyond the mere price of the plane ticket. She knows that she and her children will suffer from rising temperatures, storms, and droughts, which will have been partly caused by her flight to Mallorca. The Irish ultra-low-cost airline Ryanair is among Europe's ten biggest carbon emitters.[4] Sophia would like to avoid such damage. If she believed that it would help achieve that goal, she would be prepared to do without her weekend trip to the Spanish holiday island.

She is caught in a prisoner's dilemma. The amount her flight contributes to climate change is small, so she can argue that the personal benefit promised by the holiday outweighs the damage it causes. And, since she must assume that everyone thinks this way, she and her fellow plane passengers find themselves in the same situation as the two suspects who both know that remaining silent would bring about the best result for everyone, but who incriminate each other out of fear of being betrayed. Taking that plane is 'Nash'—even if Sophia believes that all her fellow passengers would have willingly given up the flight to help protect the climate. Just like her, they do not believe they can trust their peers to do the right thing. In fact, Sophia is in an even trickier situation than the two prisoners. At least their fate only depends on the decision of one other person, namely the other suspect. Sophia, however, would have to depend on many millions of people having the benevolence to limit their choice of transport for the sake of saving the Earth.

But how do those justified doubts fit with the firm belief expressed by survey respondents that everyone can contribute to tackling the climate crisis? Can it be that almost all Europeans overestimate their own ability to make an impact? Or do they interpret the question cynically as one about what everyone *can* do, but nobody does?

A blocked revolution

At least our holidaymaker can hope that she will enjoy her short break. She's looking forward to it. That feeling has become increasingly rare when people decide to treat themselves. There has been a fundamental change in attitudes to wealth, consumption, and living the good life in the past few decades.

This change can be seen in the results of the large-scale surveys regularly carried out among the population of Germany by social scientists. In the first series of surveys in 1980, 70 per cent of respondents admitted to having materialist values. They ranked possessions, combating inflation, and the maintenance of law and order as the most important objectives.

Those priorities have now been turned almost upside down. Fewer than 40 per cent of the representatively selected respondents now support materialist values, while 60 per cent prioritise post-materialist values such as inclusion in social decision-making, social participation, and freedom of expression.[5] A similar reorientation has taken place in almost every Western industrialised nation; the American sociologist Ronald Inglehart calls it a 'silent revolution'.[6]

Inglehart puts this change in values down to increased prosperity in those countries. The extent of the positive effect of having money depends on how much money you have. An increase in income improves poor people's life satisfaction more, because it means they can meet needs they could not meet before. Conversely, modern happiness research has shown that the improvement in life satisfaction caused by each additional dollar or euro diminishes as wealth increases.[7] And, Inglehart explains, as the marginal utility of wealth diminishes, interest in materialist values also declines.

International comparisons support his argument.

Materialist values are dominant in poor countries, while post-materialist values dominate in rich ones. Thus, people in Ethiopia and India are more likely to say that the most satisfying thing about their work is the pay, while in Sweden or New Zealand, for example, creativity and professional autonomy are seen as more important. There is a similar difference between generations. In countries with increasing prosperity, such as Germany, the younger people are, the less important they consider pay to be, and the more important they see creativity and professional autonomy as decisive factors in job satisfaction.[8] This turn towards post-materialism can be used as a lever to help reduce our use of resources, since it is the rich who cause the most environmental damage with their consumption.

So far, however, this revolution has taken place inside people's heads rather than in reality. Take a walk through any German city, and you would not exactly get the impression that people have forgotten how to spend money. Post-materialist values or not, the roads are packed with more cars than ever. There are now 72 vehicles for every 100 inhabitants in Germany. If that trend continues, there will soon be more cars than people.[9] And more generally, spending on private consumption in Germany has risen by as much as 40 per cent in real terms over the past 30 years.[10]

Young people today may be less inclined than their parents to sacrifice their lives in the hope of a good career. But how did 73 per cent of the representatively selected students answer the question in a government-commissioned survey about what counts most in life? 'The ability to afford nice things.'[11] Don't young people profess to value such things as wellbeing, creativity, and autonomy? Don't they realise that consumption is more of a hindrance than a help in achieving those aims?[12]

There is another, more probable, explanation. According to this interpretation, people *are* aware of the contradictions. They do not buy things they don't need in the belief that they are doing something good for themselves. Rather, their consumption is motivated by the need to keep up with others — all too often reluctantly. When young adults squander their precious pocket money on expensive branded clothes, when billionaires have superyachts of ever-more ludicrous dimensions built, or when ordinary citizens take on debt to buy a car befitting their status, there is always a Nash equilibrium at play. All are perfectly aware that such status symbols make no one happy, and that it would make far more sense to spend their money on other things. But as long as everyone continues to play along, no one can quit the game without losing status. Consumers are caught in a prisoner's dilemma.

A research paper published by economists at the University of Chicago calls this a 'trap'.[13] The research, carried out in 2023, yielded some striking figures. Almost every iPhone owner (91 per cent) said they would prefer Apple to release a new model every two years, rather than annually as the company does now. Almost half (44 per cent) of those who buy luxury brands say they would prefer to live in a world without Gucci, Rolex, Porsche, and Tiffany. That figure reaches a whopping 69 per cent among those who do not own any luxury goods.

The paradox of modernity

Behaviour is slower to change than values. When new insights spread through a community, opinions about what constitutes moral behaviour begin to waver. And the more liberal a society is, the greater the time lag between a change in morality and a change in behaviour.

We might expect the opposite to be the case. Liberal societies are considered to be particularly adaptable. They promise the opportunity for everyone to strive for the optimum fulfilment of their needs, free from despotism. When their needs and beliefs change, everyone is supposed to be free to alter their behaviour.

In an ideal liberal state, the government has no right to tell individuals how to live their lives, and no ruler has the right to impose objectives on the state. When citizens are no longer satisfied with a political direction, they can replace the government in a non-violent process. There is also no unalterable religious interpretation of morality. The rules of coexistence can be rewritten at any time, if it is the will of the majority. The economy is not controlled according to a predetermined plan, but by market demand. In such a free market, those providers will flourish who best meet the demands of consumers. And science provides a way to gain knowledge, whose success depends on dismantling dogmas. All theories are open to debate, and are discarded when they are refuted by new findings.

In short, we live under a trinity of liberal democracy, capitalism, and empirical science, in a system that is capable of evolving. In theory, such a society should be able to react optimally to the beliefs and desires of its members. In such a society, the clashes between private morality and public behaviour we saw in the previous chapter should not exist.

It was with precisely this aim that liberal societies developed. Disgusted by the despotic rule of their princes and repelled by the dogmatism of the church, the thinkers of the Enlightenment sought a better way of living together. They assumed that people are guided by reason and act in their own interests. They believed that if a society persists in poverty, inequality, and destruction, it must be due to the insufficient application of rationality, because people are uninformed, have not yet learned to think for themselves, or there is tyrannical suppression of the use of reason. Removing such obstacles would result in progress for all.

From today's perspective, we know that the Enlightenment thinkers were overly optimistic. The system, designed for constant improvement, is resistant to change. This sclerosis is neither a temporary phenomenon, nor can it be remedied with well-meaning appeals. Instead, the paralysis spreads inexorably because it is inherent in the system itself. 'The more developed modern societies are, the harder it becomes to change them, and the more unwieldy and unmanageable they become,' writes the German moral philosopher Hanno Sauer.[14] (It is not even necessary to look at our insufficient response to the climate crisis to see that he is right: just consider how long it takes to plan a new railway line.)

Sauer speaks of the promise of individual freedom contained in the social contract. Unlike traditional societies, which largely dictate to their members how they should behave, democracies must grant individuals 'unusually powerful rights of defence against the majority and the state's institutions'. That is true. The rules of democracy require that minorities be given a voice and that their voice be heard. And, as the next chapter explains in detail, those who defend the status quo have a much better starting position. That is why change can be relatively easily blocked by protests and legal action in a democracy.

However, the real roots of the paralysis run far deeper. Our system fails to recognise that people act against their own long-term interests, despite themselves. The grim insights of someone like John Nash were foreign to the Enlightenment thinkers. They had no idea that societies made up of free citizens can be trapped in equilibria in which the prevailing circumstances appear relatively to be the best of all worlds — and all innovation seems like a loss.

For a handful of euros

For change to be possible, it is necessary to escape such undesirable equilibria. But the hurdles are high. Are they insurmountable? Nash equilibria arise out of human coexistence, not out of the laws of nature. Balances can be shifted. Indeed, that is the aim of every political act.

Let's consider Sophia once again, who flew off to Mallorca on a minibreak despite her own convictions. She would be prepared to give up those few pleasant days on the beach if she thought it would end the climate crisis. Nobody on the plane wants global warming, but as long as it is left to each individual to accept responsibility for taking care of the environment, disaster is inevitable. Each individual believes their sacrifice will be undermined by everyone else. That eliminates any incentive for people to act in their common, long-term interest. The paradox is that this situation, which is unsatisfactory for everyone, results from the fact that everyone is free to act as they see fit. Those who claim that personal interests are best served by a maximum amount of individual freedom fail to understand John Nash — or are lying to themselves. Freedom is not the answer to everything.

What is missing is an instrument to coordinate everyone's different decisions. This could be done with a kerosene tax on flights, making them more expensive. When the costs exceed the expected benefits, a quick jaunt to Mallorca becomes less attractive. Choosing a destination that will have less impact on the climate then becomes beneficial. And a different, more planet-friendly Nash equilibrium emerges. The kerosene tax would bring Sophia a good deal closer to her primary goal, which is to leave a liveable planet behind for her children. So Sophia should therefore want such a tax.

What authority should have the right to impose such measures? As early as the 17th century, the statistician and political philosopher Thomas Hobbes outlined a similar problem to the prisoner's dilemma. He reasoned that it must be in people's interest to submit to a sovereign who enforces collective reason. In his influential work *Leviathan*, Hobbes was inclined to assign that role to a monarch. Hobbes' successors, on the other hand, argued with good reason in favour of democracy, maintaining that only the will of the majority of the population can legitimise the power of the state. They assumed that the common interests of free citizens would be enough to introduce laws that protect those interests.

Those pioneers of the liberal society overlooked the feedback effects that arise when the subjects are also the sovereign. Even when it is in their interests, citizens are reluctant to have limitations imposed on them. Since Sophia does not want a future that includes climate collapse, she should be in favour of flights being more expensive. And, since she lives in a democracy, she could give expression to that wish. On the other hand, she understandably wants to pay as little as possible for her trip. But how little does she want to pay, exactly? The previously mentioned survey of 30,000 representatively selected citizens from around Europe included the question 'Let's imagine that every citizen in your country had to pay an extra tax to fight climate change, proportional to his or her income. Let's also imagine this tax was formally proven to solve the problem of climate change within 30 years. Would you be ready to allocate to this new tax per year?'

Let's not forget that almost all the respondents answered in an earlier question that they believed rising temperatures would 'certainly' or 'probably' affect their lives and those of their children. And a majority of the same respondents agreed with the statement that climate change is the greatest challenge facing humanity this century.

So how much are citizens willing to pay for a

solution to this problem, with the proviso that the rich would have to pay more than poor people? Almost a third (30 per cent of those interviewed) rejected the idea of paying even a single cent. Another third (33 per cent) said they would be prepared to sacrifice up to 20 euros—per year! A little more than a quarter (26 per cent) said they would be willing to pay a tax of 20 to 100 euros annually. Not even a tenth (9 per cent) said they would be prepared to contribute 100 euros or more a year to save the climate.[15]

Would Sophia, as a German, be more generous? Unfortunately not. Germany may be a rich country and have a reputation for being particularly environmentally aware, but the survey showed that Germans are around the European average. There was one set of Germans that stood out as being unusually large: those who were not willing to pay a single euro—that is, those who rejected the tax outright.

The survey can be criticised for allowing respondents to name the amount that they would be prepared to pay for a solution to the problem and that would be enough to ease their conscience. Other studies avoid such a lack of clarity, usually by asking citizens to agree or disagree with suggested additional costs for energy, food, and flights, for example, to mitigate climate change.[16] Ultimately, however, it doesn't matter. All studies have come to the same conclusion: people are, at best, willing to pay about as much for a solution to the problem they see as the biggest challenge of this century as they shell out for one or two beers per week at their local pub—if at all. We would like to believe that we want to stop climate change but cannot. In truth, it's the other way around: we can do it, but we don't want to.

Not so real, please

Much has been written about this absurdly named 'willingness to pay'.[17] A more fitting name would be 'unwillingness to pay', as the phenomenon shows up in all imaginable matters: people want to live longer, but not work longer; they want to be entertained by electronic devices, but don't want any of the disadvantages of mass computerisation; they want childcare, home deliveries, and nursing care when they're old, and are not willing to do those unattractive jobs themselves, but rail against immigration that could help fill those posts.

The pattern in such surveys is always the same. When people are asked in an abstract way whether solving a particular problem will require painful decisions, the overwhelming answer is 'Yes'. No one disputes the fact that the climate crisis will require us to rethink our concept of prosperity, or that artificial intelligence will revolutionise industry. Everyone recognises that as our society ages, we will need young workers and that they will have to come from somewhere. People put out passionate appeals about how we cannot go on like this forever.

However, when empirical social researchers probe deeper with their questions about this, the result is reminiscent of Heinrich Heine's satire on the political poets of the *Vormärz* period in the early and mid-eighteen-hundreds:

> German bard! extol our glorious
> German freedom, that thy lay
> May possess our souls, and fire us,
> And to mighty deeds inspire us …
>
> Crash, kill, thunder like a devil …
> To this cause devote thy singing,
> Thy poetic efforts bringing
> To the common public's level.

As soon as specific measures are mentioned, support usually dwindles.[18] Citizens resist measures to escape the dilemma they are all trapped in. The measures would ultimately leave most people better off, but, suddenly, no one seems to want to move away from the status quo.

This is not because people are unable or unwilling to hold back their own interests for the sake of a long-term common cause. Very few people have trouble waiting until the light turns green at a pedestrian crossing, even when the road is clear. We respect the rules because they are familiar to us. We resist new restrictions — even if they are in our own interests. Why that should be so is explained by the next illusion.

Sixth Illusion: We only want the best

'And it ought to be remembered that there is nothing more difficult to take in hand, more perilous to conduct, or more uncertain in its success, than to take the lead in the introduction of a new order of things. Because the innovator has for enemies all those who have done well under the old conditions, and lukewarm defenders in those who may do well under the new.' This warning was issued by Niccolò Machiavelli, the greatest political philosopher of the Italian Renaissance, to the politicians of his time.[1] 'Thus it happens that whenever those who are hostile have the opportunity to attack, they do it like partisans, whilst the others defend lukewarmly, in such wise that the prince is endangered along with them.'

The management of the Coca-Cola Company should have read Machiavelli. It would have spared them the most embarrassing product launch of all time and one of the most expensive failed marketing campaigns in economic history. That said, the company had every reason to venture to introduce a change. Coca-Cola's market share had fallen dramatically in the early nineteen-eighties — young people were drinking Pepsi. And customer surveys left no doubt about the reason: it was the taste.

The company's directors decided to modernise Coke's recipe. Teams of food chemists, flavourists, and sensory analysts were assembled, sworn to secrecy, and sent to secret laboratories. They concocted ever new mixtures of water, sugar, black food colouring, phosphoric acid, vanillin, and caffeine. Meanwhile, market researchers invited tasters from across America and from every demographic group to blind tastings. The process took three years. Two hundred thousand people — more than half of 1 per cent of the United States' population — tested the Coca-Cola of the future. And the tasters were impressed. They rated the new Coke as far superior to the competition. It tasted fresher, sweeter, better than Pepsi — and certainly better than the old Coke. The scientists had every reason to be pleased with their results.

Only just under 10 per cent of the tasters were sceptical. Some of them said they might not stay faithful to the brand if Coca-Cola changed its recipe. Market researchers noted that this small minority tended to stir up the majority against the new taste in group-testing sessions.[2] However, the directors ignored this warning sign.

A marketing campaign was launched. On 23 April 1985, the old Coca-Cola was removed from supermarket shelves and replaced with bottles and cans bearing a 'New Coke' logo, and the dispensers in fast-food restaurants now only served the soda with the altered recipe. Television commercials, newspaper ads, and even the evening news exalted the introduction of the new fizzy drink as the dawn of a new age. All the signs indicated that the launch had been a great success. Consumers in New York, Miami, and Detroit were impressed, Coca-Cola sales recovered, and the share price went up. 'Change,' a triumphant Coca-Cola executive declared, 'is something the American people identify with.'

Everyone in the United States was talking about the new taste, but not all were in favour of it. Newspaper reporters managed to unearth a few dissatisfied Coke drinkers. A provincial paper in Wisconsin ran an article in which a 19-year-old lamented, 'So much of my life is changing outside of my control. Now Coke, the one thing left from my childhood, has been changed.' In San Antonio, Texas, a man named Dan Lauck, who was in the habit of drinking five cases of Coke a week, mourned his insurmountable loss, claiming that the day the new flavour was introduced was 'the blackest day of my life'. And in Seattle, Washington, a retired property speculator by the name of Gay Mullins formed a campaign group called 'Old Cola Drinkers of America', set up a telephone hotline for people to voice their complaints, and filed a class-action lawsuit against the Coca-Cola Company. The case was promptly dismissed. It later emerged that Mullins had tried to blackmail the company, demanding a six-figure sum to stop agitating against the drink.[3]

Events gathered pace. The company received tens of thousands of letters of complaint. Very few of them criticised the taste per se; instead, most of the letter writers lamented the loss of the beverage they knew. A psychiatrist hired by Coca-Cola to listen to the distraught calls said some of the callers were like grief-stricken parents mourning the death of their loved one. Newspaper columnist wrote fiery articles accusing the company of robbing the American people of their freedom.

After just 79 days, the company backtracked. The executives were contrite. One explained, 'The simple fact is that all the time and money and skill poured into consumer research ... could not measure or reveal the deep and abiding emotional attachment to original Coca-Cola felt by so many people.' On 11 July 1985, the company reverted to the old recipe, and has stuck by it ever since — despite the fact that the altered blend came out top in almost every blind taste test.

Chocolate and a Nobel Prize

The majority of Coke drinkers wanted the change, and only a small minority rejected it. Where did that minority get the courage to take on a global corporation and force it to bring back the original flavour?

Clearly, the problem did not lie with the new taste, but with the loss of the old one. The Old Cola Drinkers of America often repeated their complaint that the Coca-Cola Company had taken something away from them. They saw the classic soda as belonging to them, and considered themselves victims of injustice when the familiar — *their* familiar! — drink was removed from supermarket shelves.

And yet the plan had seemed so simple: the loss of the old flavour would be more than made up for by the new blend that almost everyone preferred. It should have easily returned the brand to profitability. There was no reason to expect such bitter resistance. Or so the marketing strategists thought.

Five years later, the behavioural economist Richard Thaler showed in some now-classic experiments why the gamble did not pay off. In one study in 1990, he asked his students to choose between a bar of Toblerone chocolate or a coffee mug as a reward for performing a minor task. The subjects were then randomly given one of the two rewards, as well as the chance to swap it if they wanted to. Almost no one took advantage of the option to swap. Those who had randomly received the coffee mug kept it, and those who got the chocolate bar also kept their reward, even if they had expressed a preference for the mug when questioned earlier.

This behaviour goes against the tenets of traditional neoclassical economics. Current economic theory assumes people will act according to their preferences: we want the best, and the more desirable we think something is, the more we are willing to invest in getting it. But Thaler's test subjects simply stuck with what they had, even if it wasn't what they wanted. The students didn't even have to invest anything in the simple act of swapping their reward. All they had to do was point their finger to trade with one of their peers. Instead, the students did nothing. It was as if the sweet-toothed subjects and the passionate coffee drinkers had forgotten their preference for chocolate or nicely designed mugs, preferring to stick with what they had been given. Thaler called this strange behaviour 'the endowment effect'.

These experiments were so convincing and so revolutionary that they earned Richard Thaler the Nobel Prize in Economics in 2017. A red-and-white mug bearing the logo of Cornell University, just like the ones handed out to his students, is on display at the Nobel Museum in Stockholm, alongside such exhibits as Albert Einstein's handwritten notes.

In the chapter on the second illusion, I spoke about the subtle power of habituation to gradually push our preferences towards what we are familiar with—a phenomenon that the Austrian poet Rainer Maria Rilke called 'the thinned-out loyalty of a habit that liked us, and so stayed, and never departed'. The endowment effect works differently. Firstly, it is instantaneous rather than gradual. Thaler's students were unwilling from the beginning to swap the reward they had been given. Secondly, it affects us without changing our preferences. Those who would have preferred the chocolate bar but received the coffee mug, which they then refused to swap, still expressed a preference for the chocolate in subsequent tests. Thirdly, the endowment effect can be observed in situations where no habituation is possible.

Parting is painful

A possible argument against the endowment effect is that the costs and benefits of a change cannot really be compared to one another. The new Coca-Cola, for example, offered a taste that was deemed to be better than before, but it deprived customers of the familiarity and memories they associated with the old taste. So, were those who rejected the new beverage despite liking the taste really acting against their own best interests?

And when German drivers declare they wouldn't swap their combustion-engine car for an electric vehicle for anything in the world—is that a case of an irrational attachment to a possession? Even if car lovers have to admit that a battery-driven vehicle has better acceleration, costs less to maintain, and can even be recharged for free from solar panels on the roof of their house, they still have an argument based on the incomparable sentimental value of their four- or eight-cylinder engine car. 'Electric cars are soulless,' opines the chief editor of one of Germany's major newspapers, who likes to be depicted sitting in exclusive sports cars, as he insists that is the only reason he is fighting for the survival of the petrol engine — and that it has nothing to do with the Ferrari in his garage.[4]

However, such objections are not based on reason. People will defend their possessions against better alternatives, even when the gains and losses can be clearly quantified. One example of this is when test subjects refuse to trade a reward they originally didn't want for one they did. Another example is the traders' retaining ownership of stocks even when they believe it makes no sense to keep them.

The following situation is another example. Imagine you are a passionate football supporter and your team is facing an important game. You want desperately to see the match live at the stadium, but there is so much demand for tickets that they have to be allocated through a lottery system. (As the father of a football-crazy son whose club rose so spectacularly quickly up the league table that it was promoted to the premier league, leaving it with a far-too-small ground in an East

Berlin suburb, I know what I'm talking about here.) Now you have had the unbelievable good fortune to be selected to buy a cheap, standing-room ticket. Two days before the game, those tickets start to be traded online for $200. Would you sell your ticket?

You are not so lucky when it comes to a later, equally important match. You don't win the ticket lottery, and now must decide whether to buy tickets on the 'secondary market'. Would you really be prepared to shell out $200 for a standing-room ticket?

If you answered 'no' to both questions, that's understandable—most people react that way. However, logically speaking, your decisions are inconsistent. Irrespective of whether you would have kept your ticket in the first scenario, or bought one in the second, you would be $200 out of pocket if you decided to cheer on your team in the stadium, or $200 richer if you just watched the game at home on television.

The endowment effect holds us back. People regularly demand greater compensation for giving up a possession than they would be willing to pay to acquire the same object. In one study of American basketball fans who won tickets to the annual championships in a lottery, the lucky ticket holders were willing to sell their prize for an average of $2,400. The same subjects said they would not be willing to pay more than $170 on average for a ticket to the championships.

Of course, a big sports fan who came up lucky in

the lottery could sell her ticket. Perhaps she'd like to spend the money on travel and just follow the game on TV. But that would require her to give up her ticket and any prospect of seeing the event live.

Giving something up is difficult, even if it leads to a gain. We feel the pain of losses more sharply than we feel the joy of gains—even if the object being given up and the object being gained are objectively of equal value, and often even when the change leads to a net gain for us overall. Uncertainty increases this asymmetry. And is there any such thing as a completely risk-free decision anyway?

These insights were also worthy of a Nobel Prize—this time, for the Israeli American psychologist Daniel Kahneman in 2002. His extensive research showed that people are more afraid of small losses than they are pleased by big gains. He proved that we base our decisions less on the expected final outcome than on the losses we would have to accept to reach that result.

The reason for this is that the brain generates positive and negative feelings in two completely different ways. It interprets the hope of a win as the expectation of a reward. Starting in the midbrain, this initiates processes that motivate us to act. This then activates areas of the brain responsible for muscle movement, triggers the release of hormones such as dopamine, which give us feelings of positive anticipation, and awakens our interest and willingness to learn.

The prospect of a loss, on the other hand, creates feelings of fear and rejection that can be subconscious. The amygdala, a paired structure in the temporal lobe of the cerebrum, sets the course for danger aversion. Sometimes this triggers aggression, sometimes flight, but the most common response to a threat is avoidance. As if paralysed by fear, we attempt to minimise the damage by doing nothing.[5]

Positive feelings thus help us to seize opportunities, while negative feelings help us avoid threats. This is why our reaction to a possible loss is so much stronger than our response to the prospect of an equally large gain. Missing out on a benefit has no consequences, while not reacting strongly enough to a threat can, in the worst case, result in death.

This programme for survival, after millions of years of evolution, makes itself felt every day. It is also responsible for the endowment effect. We find it so difficult to act according to our preferences because we process losses and gains so differently, and we are additionally paralysed by the fear of loss. The result is that we stick with the status quo. Individuals cling to what they have, even if they came by it more or less by chance.[6]

Sick societies

Losses due to change not only seem to us to outweigh any gains, but they are also more present in our minds.

This is due to the order in which we experience them. We almost always feel negative effects long before we see any advantages. This means we perceive costs immediately and tangibly, while the prospect of benefits feels distant and abstract. And patience is not exactly a human strength. We share this trait with other intelligent animals, which indicates that it is deeply rooted in our evolution. Dogs, rats, pigeons, crows, and apes have been observed to reject food if they can expect to receive a better snack sometime later. But they only do so in exceptional circumstances. People who lend money demand interest.

This reluctance to wait for a reward, coupled with the fear of loss, makes us shortsighted. In a flipping of the principle of interest, we accept the possibility of a worse future in order to avoid negative effects in the present. This lack of a long view is not a recent phenomenon. In his book *Sick Societies*, the American anthropologist Robert B. Edgerton analysed hundreds of field studies that refute the myth that indigenous people are perfectly adapted to the environment they live in. As Edgerton shows, in reality humans have always been prey to their own cognitive distortions.[7]

One of the many monographs quoted by Edgerton describes the lives of 'pygmies' in the jungles of the Congo. Their societies have existed for hundreds of thousands of years, and are some of the oldest in the history of the species *Homo sapiens*. In some of those societies, such as the Efe, men hunt with bows and arrows while women grow vegetables. Other societies, such as the Tswa, make use of animal traps. Among the Tswa, both men and women are involved in driving antelope into nets, which increases their rate of success. The trappers therefore regularly have more meat than they need, and can trade it profitably for vegetables.

Why don't the Efe use nets for trapping? The first field researchers suspected it was because they lived under different environmental conditions, but the rainforest and the animals it is home to are the same everywhere. In some areas, Efe and Tswa even live in close proximity to each other. Of course, the Efe are aware how much more easily their neighbours are able to provide for themselves. They even know exactly how the nets are made.

The American anthropologists Robert Bailey and Robert Aunger eventually managed to solve the mystery. The Efe declared that they, too, could undoubtedly use nets to trap animals, and it would mean more meat for them. But the nets would have to be made first, and the Efe considered that work too tiresome.[8]

When gains do not entice

Are we any different? No one actively wants to destroy the planet. Since the price of regeneratively produced energy has fallen drastically, it no longer makes economic sense to burn oil, coal, and gas for fuel. As early as 2006, a former chief economist of the World Bank, Nicholas Stern, showed in a landmark report that investing in environmentally friendly technology is profitable. The document stated that it would cost the world the equivalent of about 1 per cent of its annual economic output to phase out fossil fuels within two decades. The document added that sticking with oil and coal would be far more expensive, with the cost of the fuel itself and that of adapting to climate change likely to be between 5 per cent and 20 per cent of global economic output in size. It has since turned out that this estimate was rather conservative. Ten years after his report was published, Stern admitted to having underestimated the risks and costs of climate change.[9] Furthermore, wind power has halved in price since the report was published, while solar power is now seven times cheaper. Almost everywhere in the world, electricity produced from wind and solar power is now cheaper than that produced in conventional power stations.[10] Studies from as early as 2014 showed that price parity was achieved in Germany even before the great fall in price for regenerative energy.[11]

Any investors would be thrilled by such figures. Where else can they expect to get a 500 per cent return or more?[12] Wind turbines and solar panels make economic sense, which explains why they overtook fossil fuels for the first time in 2023 as the source of more than half of the electricity consumed in Germany. On the one hand, that success story shows that progress is possible. On the other hand, that share in the energy supply mix is still not enough for Germany to achieve its climate goals or to render the country independent of oil and gas imports. Germany lags seriously behind other countries when it comes to plans for refitting its energy-supply system. In a 2024 report, even the Federal Audit Office called urgently for more action.[13]

How can this lag be explained, given that transitioning to regenerative energy sources promises increased profits? Politicians do not reckon the same way as businesspeople do. Politicians know what concerns the people: they are less enticed by the promise of gain than they are plagued by the fear of loss.

Countries cannot avoid losses when transforming their energy supply. Jobs are lost, even though others are created elsewhere. Combustion engines and oil-fired heating systems have to be scrapped. Wind turbines and new power lines spoil the landscape. Homeowners could save a great deal of money over the years by converting their heating systems, but when they are faced with the decision to do so or not, they think only of the high initial acquisition costs.

So the mystery of why climate protection measures are easy to define but difficult to implement, even when they are economically advantageous, is explained by a complex combination of two effects.

Firstly, we perceive the costs of such a transition to be greater than the possible benefits from the outset because the human mind gives more weight to losses than to gains. The reluctance to pay for change described at the end of the previous chapter is due to the endowment effect. This creates the illusion that the current status quo is preferable to any change.

Secondly, many of the benefits of such a course correction take years, in some cases even decades, to become apparent, which further diminishes their appeal. The longer a benefit takes to manifest itself, the less attractive it seems. Conversely, the more distant the negative impact of behaviour that is pleasant now is perceived to be, the more reluctant we are to change it. The entire world is facing the same dilemma as the 'pygmies' who preferred to hunt for meagre fare with their bows and arrows than spend time making nets—both are investment-avoidance behaviours.

Why 'blood, toil, tears and sweat' doesn't work

Machiavelli's advice was only to try to implement changes that can be enforced with violence. If the prince would need to use kind words to convince his opponent, the endeavour is already lost. In that case, according to Machiavelli, it is better to leave the situation as it was.

We no longer have that option. Our world is

inherently unstable. Those who reject change will fail. But Machiavelli accurately described the dilemma posed by any reform: change entails giving up the status quo, and that incites resistance. Those who are too obvious about wanting change will also fail.

According to this argument, the Coca-Cola Company should never have publicised its change to the recipe. Machiavelli would presumably have recommended avoiding the rebellion against New Coke in the first place. Research in the wake of the fiasco did indeed show that no one would have noticed a gradual change in the drink's flavour.[14] Those who want to introduce change should keep quiet about change.

However, obfuscation is often not an option. The riskiest strategy, then, is to set out all the impacts of the change honestly. But that focuses attention on the losses, and all thoughts of the gains are gone. Eventually, the reformers find themselves with their backs to the wall: the overall positive outcome of the endeavour, which was the whole point of the change, is forgotten.

The environmental movement has had to learn that the idea of making such sacrifices is not very popular, even if they are necessary. Activists continue to call on people in wealthy countries to reduce their consumption. They take inspiration from Winston Churchill, who appealed to the people of Britain to fight for their freedom with 'blood, toil, tears and sweat'. Churchill's appeal worked because nobody in 1940 was in any doubt that the United Kingdom was facing possible destruction. The destruction of the environment, on the other hand, is happening too slowly to cause such alarm, so the activists' appeals for people to make sacrifices fail to move many. Of the rest, many fear the loss of their habitual behaviour more than the loss of the natural conditions upon which our life depends. Thus, the majority are more inclined to call the activists' analysis of the situation, and therefore also their credibility, into question than to change their own behaviour.

A more promising approach is to shift perspective and focus on the benefits of a given change. This gives hope that people will be driven by curiosity and their desire for a better life to overcome their fears. There are many positives to be highlighted by those who recognise the limits to growth and so advocate for an end to unlimited consumption. For example, when people are freed from the pursuit of money and status, their lives become healthier. They have more time to spend with their partners, family, and friends, and to pursue their passions rather than drudging away endlessly at the office. And instead of fleeing their dreary day-to-day routine several times a year by jetting off to the ends of the Earth, they can enjoy the bounty of nature outside their front door as it gradually regenerates.

On top of the pavement lies the beach

People adapt astonishingly quickly to altered circumstances. They learn from experience that their fear of change was worse than the change itself. For example, the prospect of having less money in their wallet in future makes even wealthy people worry about whether they will be able to make ends meet, and keeps them awake at night. But as soon as those losses become reality, they become less significant. This was demonstrated in studies where Americans were asked about how much tax they would be willing to pay for their state to take part in a regional carbon-mitigation policy. Respondents who believed the carbon tax was already enshrined in law and applied to everyone, including themselves, were willing to pay three times more than those who believed the tax to be a legislative proposal.[15]

Fear of loss paralyses us because our only reference point is the status quo. That means we do not judge alternatives according to their actual benefits, but according to the losses they entail compared to our current situation.

At present, most people find it difficult to imagine not being able to park their car wherever they want in the city because the roadsides where cars used to stand are now a space for cyclists to ride their bikes, children to play, and trees to grow. They are even more unsettled by the idea of closing major roads to cars entirely. And the suggestion that motorists should pay extra to drive into the city centre is seen by many as an infringement of their freedom.

Change is possible, nonetheless. While new swathes of Germany's cities' residential areas are still being asphalted over to build highways, tranquillity has returned to the banks of the River Seine in Paris. The city's governing council voted in 2016 to close the motorways that had carried traffic along both banks of the river for decades, turning them into peaceful parks instead. Since then, the *Quais* have once again been the preserve of lovers and *flâneurs*.

Congestion charges have long been a reality in London, Gothenburg, Milan, and many other European cities. They faced resistance at first. Many predicted that the charge would ruin businesses, stifle nightlife, confine old people to their homes, and make it more difficult for workers to commute to their jobs. However, when none of those fears materialised, the opponents fell silent. Detailed research documented the pace of this change in attitude. Just a few months was enough for the vast majority of citizens to be convinced of the advantages of empty roads and a better quality of life with less traffic noise and cleaner air.[16] Limiting the use of cars in the city was then seen as an acceptable price to pay.

The most powerful way to combat the fear of loss does not actually combat that fear directly. It works when we temporarily put the status quo out of our mind and mentally transport ourselves to a possible future. This is how imagination can be used to break the tyranny of habit. To imagine an improved life in a vivid, detailed, and colourful way is to overcome one of the biggest barriers to achieving that life goal. It removes the status quo from its position in our mind as the measure of all things, and allows us to assess the pros and cons of a proposed change more objectively.

Every year for 15 years before the expressways along the banks of the River Seine were permanently turned into parks, they had been temporarily closed to traffic during the summer break. Riverboats with sunken swimming pools were moored along the river, with rowing boats and exercise equipment available to all. The road surface was covered with golden sand brought from the North Sea coast free of charge by a construction company. Adapting a slogan from the 1968 protests, *Sous les pavés, la plage!* ('Beneath the pavement lies the beach'), a nearby metro station was temporarily renamed *Paris Plages* (Paris Beaches). By the end of the summer, Parisians no longer saw the reclaimed riverbanks as a utopian dream, but as the norm. Humans need to use their imagination in order to act reasonably.

Seventh Illusion: Ideologies are obsolete

Why do people see things that aren't there?[1] Solomon Asch tells of his childhood in Poland. In 1914, when he was seven years old, he was at the traditional Passover meal with the rest of his Jewish family. After pouring wine for each of his relatives, his grandmother placed another full glass on the table. The little boy asked who it was for. For the prophet Elijah, answered one of his uncles.

'Will he really drink from it?'

'Oh, yes,' said the uncle. 'You'll see.'

The child watched the glass expectantly — and was convinced he saw the level of the wine in it lower slightly, all by itself.

When the *Wehrmacht* invaded his country and the Nazis murdered millions of Jews, the family emigrated to New York. Asch later worked as a social psychologist. In 1951, he carried out a groundbreaking experiment.[2] It showed that people can infect each other with ideologies, and demonstrated how those ideologies distort their perception of reality. It also explained Asch's childhood experience.

In Asch's experiment, test subjects were asked to compare the length of some lines on a card they were given. Each card had one line at the top and three more beneath it. One of the three lower lines was the same length as the top line, one was longer, and one was shorter. The test subjects had to state which of the three lower lines was the same length as the one above. All subjects answered this simple question correctly.

Asch then divided the subjects into groups. Each group included one unsuspecting subject and seven who had received prior instructions from Asch without the knowledge of the first subject. The seven prepared subjects then unanimously gave incorrect answers, announcing them with confidence and conviction. They insisted the short lines were long and the long lines were short. And the unsuspecting subjects? They fell in with their groupmates. The same people who had previously assessed the lines correctly and without hesitation before their very eyes now declared that the lines that were just a couple of finger-widths in length to be longer than ones that ran across almost the full width of the card. Fewer than one in four subjects were able to withstand being persuaded by the nonsense they were told by the group, and stuck to the truth.

Asch put this down to a fear of deviating from

the group. In follow-up interviews, some subjects said they began doubting their own powers of perception in the face of such convincingly framed opinions from the group. Others said they had known the group was wrong, but had not wanted to make a fuss. Some subjects even went so far as to admit that they thought something must be seriously wrong with them. When their peers even saw the length of simple lines differently, these subjects felt their doubts about their sanity were confirmed.

Asch was driven by more than just a search for the quirks of human behaviour. He wanted to understand what happened in Germany during the Nazi regime. How was it possible for millions of Germans to allow themselves to be taken in by obvious propaganda lies? Even good-hearted citizens came to see their Jewish neighbours, who had never harmed a hair on anyone's head, as dangerous enemies. And even long after their cities had been reduced to rubble by Allied bombs, they continued to believe in the 'greatest general of all times', and carried on suffering in the name of ultimate victory. What had happened to the Germans' sense of reality?

Post-factual politics

The *Routledge Encyclopedia of Philosophy* defines an ideology as 'a set of ideas, beliefs and attitudes, consciously or unconsciously held, which reflects or shapes understandings or misconceptions of the social and political world. It serves to recommend, justify or endorse collective action aimed at preserving or changing political practices and institutions.' Social psychologists would add that ideological thinking is characterised by two things: firstly, it results in judgements that are biased in favour of the in-group, and, secondly, it leads us to adhere to our convictions even when they are contrary to the facts.³

For example, which of us has ever encountered someone wearing a burqa outside of Afghanistan? I haven't, although I live in Berlin and am often in parts of the city with large Muslim communities. Very rarely, perhaps once every couple of months, I might see a woman with her hair and face covered by a niqab. But I have never seen anyone wearing a burqa—that is, a full-body veil that also covers the entire face, with a small mesh screen to look through. And that is not surprising. Experts on Islam will confirm that such garments are worn exclusively in Afghanistan and some regions of Uzbekistan, Tajikistan, and Pakistan.

Despite this, after the wave of immigration in 2015 triggered by the war in Syria, so many people in Germany felt threatened by the burqa that the governing Grand Coalition at the time passed a so-called burqa law. It forbade German civil servants and military personnel from 'covering their face while performing their duties'. When a Green member of the Bundestag asked parliament exactly how big the problem with burqas was, the cabinet had no answer: 'The Federal Government has no information on the number of women who wear the burqa in Germany.' This was stated even more bluntly by the Conference of the Interior Ministers of the Federal States and Central Government in response to a question from the Bavarian public broadcaster, *Bayerische Rundfunk*, as to how many German civil servants wore a veil at work: 'We know of no such case.' Although no one knew what its actual purpose was, the 'Law on Sector-Specific Regulations for Face Coverings', came into effect in Germany on 15 July 2017.

A look at opinion polls helps to understand such instances of post-factual legislation. In a survey carried out by the Bertelsmann Foundation in 2015, 57 per cent of non-Muslim residents of Germany agreed with the statement that 'Islam is a threat'. And such people felt not just threatened, but also surrounded. When the opinion pollster Ipsos asked Germans to estimate in a 2016 survey how many of their fellow citizens were Muslims, the average guess was 21 per cent. And how many Muslims really live in Germany? Since Islam does not register membership numbers as Christian churches do, figures vary depending on the definition of Muslim used. Probably the most accurate estimate comes from the German Federal Office for Migration in 2020, which puts the percentage of the population who follow the Muslim faith at somewhere between 6.4 and 6.7 per cent.[4]

Thus, even a conservative-run intelligence service such as the Bavarian State Office for the Protection of the Constitution (its domestic intelligence service) describes the idea that Europe needs to defend itself against the uncontrolled influence of people of other faiths as part of an ideology. Demands that Europe let in as few people as possible from other cultures for its own protection, and that immigrants who already live in Europe should assimilate, fulfil all the criteria for ideological thinking described above. Those who present such arguments show bias towards their own in-group, and their claim that the country is being overrun by foreigners is not supported by the facts, since Muslims make up 7 per cent of the population at most.

People's rejection of immigration is not usually based on their personal experience—significantly, the fewer Muslim or other immigrant neighbours that Germans have, the more sceptical they are about migrants.[5] The places where the fear of migrants is voiced the loudest are those where the chances of meeting a migrant are the lowest.

It is no wonder, then, that even appeals to people's self-interest have little effect. If someone is convinced that women wearing burqas are a threat to our peaceful coexistence, it is unlikely that their minds will be changed by the fact that Germany would cease to function without the labour provided by citizens from other parts of the world. They will be equally unimpressed by economic forecasts that say Germany will need at least half a million new migrants per year if it is to maintain its current standard of prosperity.[6] Such people's fears are of a different kind. In a survey in 2024, 53 per cent of citizens expressed their 'great concern that too many people were coming to Germany'.

Burqas everywhere?

Experts are quick to expound on such fears in talk shows or newspaper columns. They declare that Germans are uncomfortable with globalisation and that they fear for the survival of their culture, or that they are worried about the high-precision mudslinging of TikTok and similar sites as they spread resentment and lies.

That may all be true, but it bypasses the real question: Why are people so susceptible to ideologies? How does someone develop an unshakeable conviction that one in every five people living in Germany is a Muslim, despite the fact that they barely know any Muslims themselves?

Asch's experiment provides an explanation for this kind of distortion. If people would rather agree with the majority opinion about the length of lines on a card than trust the evidence of their own eyes, they are also open to other suggestions. When people have been told often enough that peace in Germany is threatened by hordes of veiled women, they will probably also hold that view, even if they have never even encountered such a woman personally. Asch hypothesised that his test subjects only agreed with the prevailing opinion as a precaution, to spare themselves and the group any conflict, and actually knew that it was factually incorrect. He believed his subjects were in a state of cognitive dissonance between what they saw and what they said.

Unfortunately, Solomon Asch was overly optimistic. In 2005, neuroscientists replicated his experiment using 21st-century technology. Gergory Berns of Emory University in Atlanta and his colleagues used magnetic resonance imaging to monitor the activity in various parts of their subjects' brains as they were giving their assessments of what they saw. Rather than judging the relative length of some lines, subjects had to decide whether two images of abstract shapes represented the same figure, with one spatially rotated, or were two different figures. They were able to follow the other subjects' responses to the same task via a monitor. As in Asch's original experiment, a chorus of nonsense opinions came from the complicit subjects, and once again, the unsuspecting subjects fell in with the majority opinion. The activity measured in their brains during this process was particularly shocking. No signal was registered in the parts of the brain responsible for processing lies and contradictions. There can only be one explanation for this: the conflict between the information subjects received from their eyes and what they were told by their peers did not penetrate that level of consciousness.

Does this mean their brains were filtering out contradictions at a very early, still unconscious stage? As soon as the scientists' accomplices confused a test subject with an obviously false answer, there was a telltale burst of activity in the regions of their brain responsible for spatial perception. It seems the raw data the subjects were receiving from their eyes were being processed there to make their perception correspond to the claims they were hearing from others. In this way, no one had to be lying! No one was suffering from cognitive dissonance. The subjects really didn't see the figures as they were. They saw them as they were supposed to see them.

The brain is a forgery factory. The opinions of others can make us perceive a reality that does not exist, just as optical illusions do. This is the scary conclusion from Asch's experiment: if everyone is talking about burqas, we see burqas everywhere.

The editors of the German political magazine show *Monitor* once very commendably counted the issues discussed on the big talk shows on German television. In 2016, for example, immigration dominated everything. No fewer than 54 per cent of talk shows on Germany's public channels *ARD* and *ZDF* were about refugees, Islam, terrorism, and right-wing populism. Week after week, they showed video of figures with their faces covered, alternating with clips of enraged citizens demonstrating against them, before a panel of criminologists, politicians, and feminists argued about them. Since then, many people may be convinced of having seen covered women in their locality, too. And those people are presumably telling the truth, even if not a single woman takes to the streets in Germany wearing a burqa.

In praise of simple-mindedness

We believe apes simply mimic behaviour, but credit *Homo sapiens* with a critical consciousness. However, recent experiments, in which our social intelligence was compared to that of chimpanzees, cast some doubt on that idea. The results show how we do not perceive what exists, but what our peers tell us exists.

In a fascinating experiment in 2007, scientists at the University of St Andrews in Scotland gave the same task to some preschool children and some chimpanzees. They had to open a tube to get to a reward of fruit inside. The tube had various locks and levers, all of which were useless except one. The researchers demonstrated a complicated (but pointless) sequence of manipulations to the children and apes, which they proceeded to mimic. As soon as the chimpanzees realised that they only had to press one lever to open the tube, they ignored all the rest. The children, on the other hand, stubbornly continued to try to use the unnecessarily complicated procedure they had learned from the researchers, even when they had seen that there was an easier way.

The psychologists in Scotland called this stubborn behaviour 'overemulation', and it is by no means confined to children. Adults are even more persistent in emulating the absurd behaviour of their role models. This has also been demonstrated in experiments. Extreme conformism is not peculiar to our culture. Even hunter-gatherers in South Africa living in largely Stone Age conditions feel compelled to copy what they see. All this indicates that humans have a strong inborn urge to happily imitate what we see before us.

To understand any genetically programmed trait, it is helpful to take an evolutionary view. There must be reasons why humans have a far greater propensity than any other animals to emulate the behaviour of others. Joseph Henrich, an anthropologist at Harvard University, even believes this trait to be the reason for the success of our species.[7] He argues that cultures were able to come about only because our brains absorb everything our fellow humans say and do like sponges. If we were as naturally sceptical as chimpanzees, we would never have learned to use fire, writing, or smartphones. Learning cultural skills requires that we adopt habits even if we don't understand the sense of them.

Cultures make life complicated. Questioning everything is dangerous. The anthropologist Henrich gives the example of manioc, a root vegetable that has been a staple in South America since time immemorial. Unless processed, raw manioc can release toxic hydrogen cyanide. Women in the Amazonian lowlands peel and grate the tubers, and then soften the pulp by boiling it several times until it is no longer toxic. The women spend more than a quarter of their time on this process without knowing why they do it. They simply do as their mothers and grandmothers taught them. And, because everyone always does it, none of the women has ever seen a case of cyanide poisoning.

If those women were chimpanzees, they would only do what seemed understandable and necessary—removing the bitter taste from the manioc. That can be done by simply boiling the root vegetable. However, traces of hydrogen cyanide would remain in the manioc, and would gradually kill those who ate it. Symptoms often take years to appear. They can include weakness, paralysis, and developmental disorders in children. However, the women in Columbia have an inborn propensity for overemulation. They repeat exactly what they have been taught without even thinking about it. That overemulation saves their lives — and provides the people of South America with one of their most important staple foodstuffs. Imitation and acceptance without question translates into reproductive success.

Overemulation only works if we are prepared to accept things that seem absurd to us. Without this ability to trust blindly, human culture would never have emerged. Once ideas are fixed in people's minds, they are passed down from generation to generation. Evolution is not concerned with objective reality; the chances of survival are its currency. For our ancestors, being simple-minded and easily led was more useful for survival than being critical.

How we let ourselves be led

Ideologies are the dark side of any civilisation. And yet people are prepared, to varying degrees, to accept that hierarchies are immutable, or to believe in all-powerful gods, or in the blessings of unfettered markets, or in a socialist paradise, or in the good intentions of a strong leader, or in the uniqueness of their own nation. But what determines whether a person is seduced by an ideology or not?

In 2016, two major, if unplanned, experiments took place that go some way towards answering that question. On 23 June of that year, the British electorate voted to leave the European Union. That decision was based on ideology; the proponents of Brexit promised the country would return to its former greatness, although they were not able to say how leaving the EU would improve the lives of people in England, Scotland, Wales, and Northern Ireland. Experts had almost unanimously issued urgent warnings against the move, substantiating their advice with undeniable facts. They said Brexit would pose a serious threat to the stability of the country and the prosperity of its citizens. Despite that, the referendum ended in a narrow decision to leave the EU.

In November of the same year, the people of United States mystified the world by believing that a man who even then was embroiled in no fewer than 3,500 private and business court cases would act in the best interests of the nation as president. Just like the Brexiteers, Donald Trump presented himself as a bulwark against change. He promised America would be great again under his leadership, and that everything would return to the way it used to be, only better.

Journalists and political scientists usually look to people's living conditions for the roots of attitudes that subsequently lead to political change. A noticeably large number of people in the depressed former coal-mining regions of Northern England were in favour of leaving the EU. And voters who helped Trump become president proudly reclaimed the word 'hillbilly' as a badge of honour. It was believed Trump's rhetoric soothed the souls of poor, mainly white men in provincial America.

However, those plausible-seeming explanations are not borne out by the facts. Even in Northern England, 40 per cent of citizens or more were against Brexit in almost every constituency. The majority of Brexit-supporters were in the wealthy south of England and were far from being part of a disaffected working class. Around 60 per cent of votes in favour of Brexit were cast by members of the middle class.[8] The percentage of Trump voters with above-average incomes was just as high. Forty per cent of the votes cast for the New York property developer who has been accused more than a dozen times of rape and sexual assault were from women.[9]

Donald Trump's re-election as president in 2024, despite having received a criminal conviction in the meantime, is further confirmation of this.[10] Once again, the majority of Trump voters were above-average earners, although he succeeded in winning more votes from lower-income groups as well. There could be no more talk of Trump being the candidate of impoverished rural Americans; he made his biggest gains compared to 2016 in the wealthy metropolitan suburbs. He also gained more votes from female voters than ever before, with 46 per cent of women voting to re-elect him. Among white women, Trump supporters were even in the majority.

But why? As early as the nineteen-fifties, the German philosopher Theodor Adorno theorised that, alongside economic factors, there also must be personal reasons for the attraction that some people feel to immutable worldviews, while others seem largely immune to the allure of great promises. 'Ideologies have different degrees of appeal for different individuals', h wrote in his work *The Authoritarian Personality*, published in 1950, very soon after the end of the Nazi regime.[11]

Later work confirmed this theory. Moreover, large-scale studies of more than 12,000 pairs of twins carried out over four decades in five different countries showed that inclinations towards particular political views are genetic. Identical twins are not only very similar in appearance, but they are also far more likely than regular siblings to share a similar worldview. There is no such phenomenon among fraternal twins. All the pairs of twins studied grew up together in the same household, and generally attended the same school. That means the reason must lie in the fact that identical twins share the same genetic make-up. Of course, even twins sometimes argue about politics, but they do so far less frequently than regular siblings. Studies indicate that a good 50 per cent of a person's ideological views can be put down to genetics.[12]

Ideologies are a crutch

The Cambridge University neuroscientist Leor Zmigrod took advantage of the heightened tensions during the Brexit referendum in the UK and after the presidential elections in the US to find out what kind of person is susceptible to ideologies.[13] She carried out a large battery of psychological tests on hundreds of representatively selected citizens on both sides of the Atlantic. Participants in one of her studies had to undergo 22 personality tests and to complete 37 perception and reasoning tasks. This enabled her to create personality profiles of them, measure their reaction speeds, and assess their cognitive flexibility.

Participants were also tested for their need for certainty and security, and about their habits. They were asked if they found unforeseen situations worrying or exciting, and if they saw routine as pleasant and reassuring, or boring. None of the questions wad obviously about politics.

The researchers then questioned the volunteers about their voting decisions and their views on the general state of the world. Zmigrod and her colleagues asked about party affiliation, nationalist sentiments, opinions on immigration, and attitudes to authoritarianism.

The scientists were astounded by the correlations they discovered. One of the psychological tests they used took the form of a simple card game in which the rules would suddenly change without the player's prior knowledge. The test measures a person's ability to adapt to altered conditions. Another of Zmigrod's tests looked at creativity. The task was a word-association game in which the aim was to find the word that linked three other, seemingly random words, such as 'cottage', 'Swiss', and 'cake'.[14] Those who found both tasks difficult turned out to have a general inclination towards dogmatism — as if they were trying to make up for their slow thinking with answers drawn from a rigid worldview.

This phenomenon was recorded among adherents of both left-wing and right-wing ideologies, but there were differences between the two. Conservative subjects generally showed less cognitive flexibility than those with left-leaning politics. The researchers also noticed that conservative subjects had particular trouble with the word-association task. They appeared to have a very narrow way of thinking. Meanwhile, those who had trouble with the card game and its unpredictably changing rules were noticeably more probable to have a liking for authority. Also, the more that a subject appreciated routines, the more likely they were to believe in authority figures and to hold not only authoritarian and conservative views, but also nationalist ones. It is not surprising, then, that Trump supporters say they want to return to what they know. In surveys, more than 60 per cent of them agreed with the statement that America's best days were in the past.[15]

Even perception tests involving recognising innocuous patterns of dots were predictors of political orientation. Subjects who were a couple of milliseconds slower to react than average had an inclination towards dogmatism. Those who only answered when the pattern was clearly recognisable, and otherwise held back, were more likely to espouse conservative views. Those who were slow to react in all cases but did answer even when the patterns were unclear were noticeably more likely to think along authoritarian lines.[16]

Overall, the brain scientists' tests were five times better than the usual factors such as sex, age, income, skin colour, or geographical location at predicting whether someone had voted for Brexit or adhered to a worldview typical of Trump supporters.[17] They offer stark confirmation of how deeply the basic mechanisms of perception and cognition affect which convictions we hold.

Fear fosters conservatism

The reluctance to accept change and susceptibility to ideologies is rooted deeply in the brain. That is a sobering realisation for all those who believe that fact-based arguments will prevail in a democracy. The worldviews that affect our behaviour do not bow to logical arguments. They appeal to us because they are economical for our neuronal data-processing system to handle. Cognitive flexibility is the crucial trait. The more difficult a person finds it to adapt to new situations, the less able they are to question their own beliefs.

It might seem that the obvious course of action would be to train people's cognitive flexibility. But research gives cause for doubt. Although people with low cognitive flexibility are able to learn strategies to solve specific problems, they have difficulty applying those learned strategies to new and unfamiliar situations.[18] Such people have to seek the familiar whenever possible. And they stick to their convictions once they have formed them.

A person's environment can often increase the appeal of ideologies. When people perceive a threat, or are simply frightened by the theoretical possibility of a threat, they become more susceptible to worldviews that offer safety and security. When the brain is stressed, it automatically goes into survival mode, which favours rapid decision-making and minimises mental effort.[19] A survey of more than 60,000 participants on five continents confirmed this effect.[20] Whether in Germany or the United States, Israel or Russia, Pakistan or Ghana, citizens everywhere who are threatened by violence, war, or terrorism are not only more willing to submit to authoritarian leaders, but they are also apt to justify economic inequality, condemn abortion and homosexuality, and believe that men have more of a right to a good job than women.[21]

And, finally, worldviews create fertile ground for further ideologies by manipulating both our perceptions and the mechanisms behind them. All it takes is for people to read a text that begins with the statement 'I believe that there are hidden forces of nature that people do not understand yet, and that determine many important events in life.' After reading the statement, even test subjects who professed not to believe in such nonsense were more likely to see non-random patterns in abstract modern artworks and to see conspiracies behind normal, everyday occurrences.[22] The ominous sentence prompts the brain to search for patterns. Once that process has been set in motion, we begin to see connections that don't exist. And the propensity for paranoia continues to grow.

Worldviews do not only reflect the way our brains work, but they also control it.[23] Hannah Arendt's statement about the mindset of fascism is true of all forms of ideological thinking: 'What totalitarian ideologies therefore aim at is not the transformation of the outside world ... but the transformation of human nature itself.'[24]

Flex your (cognitive) flexibility

Ideologies are like the common cold, but more dangerous. Both are contagious. Neither viruses nor ideologies can be entirely eradicated from the world, but we do have some control over whether they infect us or not.

The American economist Bryan Caplan came up with an entertaining exercise aimed at breaking down inflexible thinking.[25] It is a game in which opposing thinkers switch sides and take on the role of their ideological opponents. They then have to formulate an argument for the opposite opinion to that which they actually hold, and present it to someone who *is* of that opinion. If the listener fails to notice that the speaker is only playing a role, the speaker wins the game.

This role-playing game is based on a famous exercise designed to test artificial intelligence. If a computer is able to maintain a conversation with a human without the human realising she is interacting with a machine, then the computer can be said to have developed human-equivalent powers of cognition. It was thought up by the British mathematician Alan Turing in the early days of computing. In his honour, Caplan named his game the 'ideological Turing test'.

If you're the kind of person who has been active in refugee-aid charities for years, try setting out why Europe cannot possibly accept even more people from foreign cultures. Or, by contrast, if you believe that 'migration is the mother of all problems', as a former German interior minister once put it, try exploring your humanitarian side and explain that it is the moral duty of—and in the interests of—a rich country such as Germany to provide asylum for people who are in need through no fault of their own. (If you like, try, as a test of your own mental flexibility, 'The test of a first-rate intelligence is the ability to hold two opposed ideas in mind at the same time and still retain the ability to function,' as F. Scott Fitzgerald once famously wrote.[26])

To pass the test, players have to be able to present an argument that actually goes against their own views, credibly and without even a hint of malice. The aim is not for players to change their stance, or to simply parrot the other person's opinion. The point of the exercise is for players to realise how they deal with other people's opinions. Are they so good at understanding other people's points of view that they can present them as justified and appropriate? Or do they see opinions that differ starkly from their own as expressions of narrow-mindedness, selfishness, or stupidity? If the latter is the case, they are probably ideological thinkers: reacting emotionally saves them the effort of having to consider unusual perspectives.

It may be the case that some people who are afraid of being overrun by foreigners are selfish and narrow-minded. By the same token, it is perhaps not an exaggeration to describe some activists as hypocrites whose main motive for supposedly helping others is to shine a positive light on themselves.

However, it is highly unlikely that people would describe their own motives that way. The motives that you attribute to someone say more about your worldview than theirs. And unless you can put yourself in other people's shoes, you will have little chance of changing their minds.

Part III
How Change Works

Escaping addiction

Despite everything, times change. Anyone who doubts this should just watch a few old movies. Whether they are thrillers, comedies, or Westerns, and no matter which decade they were made in, one thing strikes the viewer immediately: the ubiquitous clouds of smoke.

Sean Connery as 007 lights up a cigarette at the casino before he introduces himself with the line that has gone down in film history: 'Bond. James Bond.' Clint Eastwood constantly tokes on a small cigar as he mows down his adversaries in *The Good, the Bad and the Ugly*, and when he finds a mortally wounded soldier in the ruins of a shot-up house, he puts a cigar between the dying man's lips. Against a swell of melodramatic music, the man inhales his final hit of nicotine with his last breath. Jean Seberg and Jean-Paul Belmondo puff incessantly as they try to evade the Paris police—the director Jean-Luc Goddard even gave the film the telling title *Breathless*. And, of course, Uma Thurman is shown with a cigarette between her fingers before taking to the stage for the famous dance scene in *Pulp Fiction*.

In this respect, Hollywood was not unrealistic. People used to smoke everywhere. Offices, meeting rooms, and university lecture halls hung with smoke; on planes, clouds from the smoking section would waft through the entire cabin, and smoking in doctors' offices was not legally banned in Germany until 2007. Doctors would even sometimes indulge in a cigarette during consultations.

No one seemed to be bothered by this. Tobacco was part of everyday life. On arriving at someone's house, it was not long before visitors would ask for an ashtray—if they hadn't already been offered a welcoming cigarette by their host on arrival. And Germany's foremost etiquette manual recommended that such an offer should be accepted without fail: 'A cigarette offered by a host should be accepted with polite thanks, and lit immediately.'[1] The *A-B-C of Good Manners for Professional and Everyday Life*, published in 1956, allowed only one exception to this rule: 'However, if you are an active sports enthusiast and truly do not smoke, you need not force yourself to do so in company.' However, hosts should not generally expect such a reaction, the guide continued, since 'these days, it is rare to encounter, even among young people, someone refusing the offer of a cigarette with the words "No thank you, I don't smoke"!'.

That's how it was. In the early nineteen-fifties, 88 per cent of men and 21 per cent of women over the age of 14 in Germany were smokers.[2] (In the UK, the percentages were similar for men, but double for women.[3]) And 'smoking' meant puffing practically non-stop. The average per capita consumption was just under one pack a day, which means most smokers must have been chain-smoking their cigarettes. Today, smokers are a minority, at just 22 per cent of men and 15 per cent of women.[4] Per-capita consumption of cigarettes has also been falling for years.

Is there anyone who longs for a return to the old days, when clothes had to be aired after every evening at the pub, and kisses tasted like licking an old ashtray?

A farewell to tobacco

Currently, the environmental crisis, demographic change, and computerisation are forcing people to change their habits. Saying farewell to tobacco was a far bigger ask of smokers. Tobacco consumption is not just a habit that raises the user's mood and banishes boredom for a few minutes at the strike of a match. Nicotine is one of the most potent addictive substances in existence. It changes the brains of those who consume it. For almost everyone, giving up nicotine after years of addiction means battling physical withdrawal symptoms, emotional distress, and the constant temptation to relapse. As Mark Twain once sarcastically quipped, 'Giving up smoking is easy ... I've done it hundreds of times.'

Escaping addiction is a battle against the body's own biology. And trying to put an end to tobacco's ubiquitous presence in society made enemies of the tobacco giants. The cigarette industry will stop at nothing to protect its profits. A list of the tactics employed by the big tobacco companies, and still used by them today, was drawn up and circulated by the US Department of Health. It includes everything from classic lobbying and financial support for favourable politicians to the targeted intimidation of detractors; from advertising campaigns and payments to tobacco-friendly scientists, to funding supposedly scientific institutes whose purpose is to sow doubt about the dangers of nicotine and tar; and from organising campaigns for smokers' rights to donating generously to relevant charities. The paper lists and gives examples of every conceivable method used to manipulate the public.[5] The four private companies that dominate the global tobacco market spent, and still spend, billions on keeping smokers addicted and on encouraging young people to get hooked.

Such tactics are strikingly like those employed later by big energy companies to block or delay the phasing out of fossil fuels. When awareness of the dangers of climate change began to grow in the nineteen-eighties,

Shell, Saudi Aramco, and ExxonMobil were already well prepared to prevent the long-overdue change. They had studied the battle plans of the tobacco industry, and learned from them.

The demise of smoking took place in less than a generation. Hundreds of millions of women and men around the world managed to escape their addiction. Never had so many people changed their behaviour in so short a time. Non-smokers also changed their way of thinking. Society developed a new understanding of the health risks associated with tobacco consumption. New social norms were established. How was it possible for such a change to succeed in the face of such massive resistance?

How to persuade women to poison themselves

Ever since the birth of the tobacco industry, one of its central aims has been to influence people. No one has a natural need to inhale nicotine and get a sore throat as a reward. The companies had to create their own consumer base. The triumph of the cigarette is a prime example of the way that a desire for things and habits can develop out of nowhere, and how those things and habits can come to be seen as natural and even essential.

Tobacco was consumed by Native Americans long before it was brought to Europe by Christopher Columbus. In the Old World, it was initially an expensive, occasional luxury for the wealthy, who took their tobacco in the form of pipe-smoking, snuff, and, later, hand-rolled cigars. It was only after 1880, when an American called James Bonsack invented a machine that could produce 12,000 cigarettes an hour using cheaper, low-grade tobacco, that the masses could afford to smoke. The next task was to manufacture a desire.

Bonsack's financial backers pulled out all the stops. They distributed 400,000 folding chairs with the Cameo cigarette brand logo to all parts of the United States, had huge advertising murals painted, placed newspaper ads claiming their product could calm smokers' nerves, and even recommended smoking as a remedy for asthma.[6] Packs included trading cards showing scantily clad ladies to appeal to young male consumers.[7]

From that beginning in America, the expansion ran its course. Around the turn of the century, the company set up to develop the Bonsack machine became a corporation that grew to dominate the cigarette market—just as John Rockefeller's almost contemporaneous Standard Oil Company dominated the petroleum business. The tobacco and oil industries were to continue to use astonishingly similar strategies to secure their profits.

The final breakthrough came with World War I, when every American soldier received a daily ration of Camels or other cigarettes. 'You ask me what we need to win this war,' said the US army general John Pershing at the time. 'I answer tobacco as much as bullets.'[8] Newspaper ads appeared, soliciting donations from the public to buy more tobacco for the troops. Consumption boomed. When the war was over, people in the ruins of Europe used cigarettes to suppress their hunger. Germany became the world's biggest tobacco importer. The industry did all it could to get even more people to take up the habit, and, once again, the new marketing methods were imported from America. Women were identified as the next target group. Advertisements praised the slimming effects of smoking: 'Reach for a Lucky instead of a sweet.' The campaign worked, but it did not change the convention that women should only smoke behind closed doors. Women smoking in public was frowned upon.

In the week leading up to Easter 1929, a public relations agency recruited women who, 'while they should be good looking, they should not look too model-y'. The idea was to pay the women to smoke Lucky Strikes as they marched in New York's traditional Easter Sunday Parade and be captured by the lenses of photographers who had also been paid by the agency to document the action. The call to join the march was written by the famous women's rights campaigner Ruth Hale: 'Women! Light another torch of freedom!'

The campaign was run by Edward Bernays, a nephew

of Sigmund Freud. He had emigrated from Vienna to New York with his family, and became known as a pioneer in the field of propaganda, which he rebranded as 'public relations'. Bernays placed great importance on underpinning his campaigns with science. His aim was to harness the power of the subconscious for economic use. Referring to him as 'Uncle Siggi', Bernays admired Freud, from whom he had learned that people rarely act out of personal conviction, despite believing they are doing precisely that. Bernays also suspected that people do not necessarily follow their personal preferences, but are far more motivated by a need to conform to a social norm, even if it is not in their own interest to do so. Thus his propaganda anticipated many of the scientific findings I discussed in the previous chapters of this book.

Bernays' analysis of the situation was as follows: women didn't smoke in public, even if they wanted to, because other women didn't do it. However, that very inclination towards overemulation could also make them reach for a cigarette even when they didn't want to. All it would take was to persuade them that other women did smoke in public. At some point it would become normal for women to light up during a date or even just while waiting for the bus. The new pattern of behaviour would then automatically become the norm.

'If we understand the mechanism and motives of groupthink, it will be possible to control and direct the masses according to our will, without their knowledge,' wrote Bernays. He called his process 'the engineering of consent'.[9]

So the increasing appearance of smoking in movies was no coincidence; it was Bernays' work. Hollywood became the second major target in the campaign to make smoking socially acceptable. Movie stars were paid huge sums to light up on screen, and scriptwriters and producers received generous fees to include scenes of smoking in their films. Writing anonymously, Bernays published a handbook on the dramatic possibilities of smoking. In it, he wrote that a cigarette can say more than many words, adding that, in the hand or mouth of a capable actor, they can express anything, from the brightest comedy to the darkest drama.

Within just a few years of that Easter parade, the proportion of American women who were smokers had leapt from 5 to almost 20 per cent.[10] But Bernays' own wife does not appear to have been one of them. As a prominent feminist, who was the first female citizen of the United States to keep her own name after marrying, Doris Fleischman would presumably have liked to be seen holding a cigarette, but no photos exist in which she is shown smoking. It is said that her husband discouraged her from the habit. Bernays was fully aware of the dangers of nicotine addiction.

Doctors smoke Camels

Lung cancer, previously a little-known disease, started to spread. As early as 1912, researchers started to suspect that tobacco smoke was the cause. In 1930, a German doctor from Saxony named Fritz Lickint published statistics that confirmed those suspicions, and American scientists soon produced similar figures. The numbers showed the doctors' worst fears to be true.[11]

Most doctors were unconcerned, however. 'We physicians of the older generation who have seen the smoking of cigarettes grow from what seemed scarcely more than a toy into what is now one of the most significant of social institutions,' wrote the prominent New York doctor and historian of medicine James Walsh in 1937, '… are under an obligation to the rising generation to warn them of the serious dangers associated with the abuse of cigarettes in our day.'[12] But he also admitted to being a smoker himself. Only 'excessive' cigarette smoking was dangerous, according to Walsh. But what counted as 'excessive'? Dr Walsh also attested that many doctors he knew smoked 20 to 30 cigarettes a day and were 'as healthy as the proverbial trout'.

Walsh fell prey to the typical illusions outlined in the previous chapters. He ignored any warning signs that did not fit with his beliefs. When presented with statistical evidence showing a rise in lung cancer cases, he lapsed into anecdotalism. Stories fascinate the mind far more than data, as we have previously seen. In the first half of the 20th century, doctors were not yet used to thinking in terms of probabilities. They believed organisms worked like clockwork automata, which meant pathogens would have the same effect on all human beings. They did not have the intellectual tools to understand correctly the information they were receiving.

There were also more prosaic reasons to avoid challenging smoking in general. Doctors were afraid of angering their patients.[13] Also, getting through a hard day at the surgery without smoking themselves was not an inviting prospect. And doctors were constantly courted by the tobacco companies. Free samples were distributed generously at medical congresses, and even as late as 1954, the country's most respected medical publication, *The Journal of the American Medical Association*, still accepted advertising for tobacco products.

White coats were everywhere in cigarette advertising for the general public. For many years, the industry employed real or alleged medics to proclaim that smoking was safe—as long as consumers chose the right brand, of course. Newspaper ads proudly proclaimed that 'More doctors smoke Camels than any other cigarette!' A full-page image of a young woman sporting a head mirror, representing a doctor specialising in diseases of the throat, assured readers that female medics were also loyal to the brand—with 7,250 of her fellow women doctors 'on the path of human service' preferring to smoke Camels.

Ads run by competing brands claimed that doctors prescribed Philip Morris cigarettes to smokers as a remedy for a sore throat. A paper published in a reputable specialist journal, one ad claimed, proved this resounding therapeutic success: after patients switched brands, all their symptoms disappeared.

The illusion that misfortune always strikes others earned the big companies billions in profits. In the nineteen-fifties, when word began to spread about how toxic nicotine is for humans, US tobacco companies commissioned several studies to find out whether people saw cigarettes as harmful.[14] The proportion of respondents who answered the question with 'Yes' was consistently in excess of two-thirds. Even a majority of smokers agreed.

The next question was whether respondents believed that certain cigarette brands were healthier than others, or were even completely safe. Almost all the non-smokers found the idea absurd. But the pattern of responses from smokers was very different. Three-quarters of nicotine addicts named a brand—mostly the one they smoked.

A storm blows over

The turning point came with a report by the United States Surgeon General, presented to hundreds of journalists in Washington, D.C. on 11 January 1964. The government chose to release the report on a Saturday morning for fear it might trigger a crash on Wall Street. The document brought together everything that was known at the time about the harmful effects of nicotine and tar, from oesophageal tumours to low birth weight. The evidence was overwhelming. As expected, the story was covered by all the media. By February, the number of citizens who believed that smoking caused lung cancer had risen by 15 per cent. Rarely before had so many people changed their views so quickly. Hundreds of thousands of smokers tried to quit. Support for 'drastic government action' to combat smoking also increased.[15]

However, by springtime, the storm that appeared to threaten the tobacco companies seemed to have blown over. Most of the smokers who had tried to kick the habit had failed. Support for anti-smoking legislation melted away. PR agencies in the pay of the tobacco industry warned of curbs on people's personal freedom, appealed to the individual responsibility of adults to choose their own behaviour, and issued reminders that, in a democracy, everyone has the right to treat their own body as they choose. The federal government, which had published the report, did nothing. Most members of the medical profession were unimpressed by the contents of the report, with its depictions of tar and nicotine as deadly poisons. Patients were often simply advised not to smoke too much.

The five phases of revolution

The philosopher and historian Kwame Anthony Appiah has described the five phases of any moral revolution.[16] In the first phase, knowledge of a problem gradually spreads, although it is not yet recognised as such. The cause of this myopia was the subject of the previous chapter: our habits distort our perceptions and shape our thinking. This was how it was possible to deny the dangers of smoking for decades, even as they were becoming increasingly undeniable.

In the second phase, increasing numbers of people begin to recognise the problem as such, but they continue to deny that they are affected by it. The government report on the devastating health effects of smoking brought society to this stage. After the press conference in January 1964 and the ensuing media attention, the risks were public knowledge. Millions of people were profoundly shocked. But rather than prompting a change, the knowledge left people nonplussed. There was a total lack of strategies to bring about change, and this was the payback for not taking the dangers of smoking seriously. People had no way to counter the combined power of propaganda and habit. They escaped the resulting cognitive dissonance by hoping that nicotine would poison others, but not them.

The third phase is resignation. It begins when cognitive dissonance breaks down. People now accept that they are affected, but continue to feel powerless to do anything about it. This stage was reached in 1970. The first generation to have grown up with cigarettes was now getting old. More and more people began to experience firsthand what they had otherwise only known from the newspapers or from television reports. Even those who were not directly affected had a family member or friend who was harmed by nicotine. Doctors were seeing increasing cases of tar-damaged lungs, smoker's leg, and coronary heart disease. Companies were losing workers. The threat had now become tangible, and the victims had faces.

People openly bemoaned the situation, which was now undeniable. However, it was still easier and less painful to find reasons not to give up nicotine than to quit smoking. Smokers who had previously celebrated the cigarette now began to express hatred for their own addiction. The use of that term was new in this context. Until the early nineteen-seventies, smoking had been seen merely as a habit rather than an addiction.

The tobacco industry reinforced the idea that individuals and society as a whole were helplessly at the mercy of the leaf. Chemists manipulated the nicotine content of tobacco to make it more addictive, and psychologists designed advertising campaigns to win back quitting smokers.[17] Lobbyists worked on politicians, and lawyers sued anti-smoking activists. Companies set up charitable funds that donated money to minority groups and women's rights initiatives, supported schools, and promoted the arts—all in order to gain as broad a public support base as possible for tobacco consumption. PR agencies ran advertisements casting doubt on the science, and paid questionable professors to publicly attack their colleagues. Company boards, along with workers' unions, never tired of pointing out how many jobs and livelihoods depended on cigarette manufacturing.

Although no one could now seriously deny that smoking was claiming hundreds of thousands of lives a year, cigarette sales remained constant. Politicians stood on the sidelines and did nothing. Whenever dedicated parliamentarians proposed effective advertising bans, or at least an increase in tobacco duties, lobbyists always made sure those initiatives came to nothing.

They had an easy job, helped as they were by the fear of loss discussed earlier. Smoking bans were cast as an infringement on smokers' individual freedoms. Advertising bans were portrayed as cutting into the revenues of media companies and agencies. Tax increases were portrayed as a threat to jobs in the tobacco industry.

But shouldn't the undeniable benefit of saving hundreds of thousands of lives a year have counted for so much more? The politicians' answer was: No. The American political scientist Thomas Marshall analysed 59 US legislative proposals to reduce cigarette smoking.[18] There were opinion polls on all such initiatives between 1964 and 2010 — mostly commissioned by the tobacco companies. The idea that people felt those draft federal laws would rob them of their rights of self-determination turned out to be false. Such proposals regularly received majority support among the respondents in those nationwide surveys. Seventy per cent of the legislative proposals were approved of by citizens, and only 27 per cent were rejected.

Politicians continued to follow the logic that the interests of those who stand to lose due to an introduced change carry more weight than the interests of those who stand to gain. So most of the initiatives failed when the cigarette industry unanimously opposed them. British American Tobacco, Philip Morris, and the rest were seldom in disagreement with each other. When they were, legislation that was supported by a majority of the population had a good chance of being adopted. But if the support of a majority of citizens was lacking, most laws failed to get through.

Using smoking as an example, Marshall demonstrated the two conditions necessary for social change: those who stand to gain from the change must want it; and those who stand to lose must not oppose it. The same pattern holds true for other efforts to introduce popular reforms—in social policy, tax legislation, or energy supply, for example.[19] Even if the proponents of a change have better arguments and more pressing interests, they will normally be unable to compete against a united front of objectors.

David versus Goliath

Given all of the above, how was it possible to banish cigarettes from public life?

The initiative came from a woman called Betty Carnes. In 1966, she lost a friend to the effects of tobacco addiction, as she related subsequently in a television interview. Her friend was only 29 years old, and the mother of two small children. Shortly before she died, her friend hugged Carnes and said, 'Betty, make people aware. Make people aware of the dangers of tobacco.' Carnes thought, 'If I had only said something, if I had only been aware and had done something then, it might not have happened ...'[20]

Carnes gave up her work as an ornithologist and set up a citizens' action group in Scottsdale, Arizona. It is difficult to imagine a less likely starting place for a global revolution than this small city close to the US–Mexican border. For Carnes, the place was irrelevant; it was simply where she happened to live. She and her action group were to change the world. But was that her aim all along?

Thank you for not smoking. The slogan was Carnes' idea. Her team of volunteers put up such signs in restaurants and businesses around Scottsdale. Business owners who refused to display the sign received a telephone call from Carnes or one of her volunteers. She realised the polite formulation of the signs might antagonise some smokers. But that was not her aim. She was not campaigning for a ban on cigarettes, nor did she want to demonise smokers. She had no ideology to preach. All she wanted was for people not to be forced to inhale smoke from their surroundings against their will. So she slightly altered the wording of the slogan to *Thank you for not smoking here.*

The signs started to appear all around Scottsdale. Bars set up non-smoking areas, and increasing numbers of businesses placed ash urns outside their entrances and declared themselves to be non-smoking spaces. That marked the beginning of the fourth of the five phases of every moral revolution, as described by the philosopher Appiah: action.

Betty Carnes' tireless action soon bore fruit. After enduring the smoke of a chain-smoking passenger seated next to her on a flight to Texas, she and her fellow campaigners bombarded the airline with letters until, in 1971, it set up non-smoking areas in its planes for the first time in aviation history. Carnes' next target was the Arizona state legislature. An initial attempt to ban smoking in public buildings failed, as the legislation did not get enough support from members. But Carnes was not deterred by the defeat, or by the tobacco companies' attempts to vilify her. A brochure published by Reynolds, the tobacco company that produces Camel cigarettes, took aim at her private life, describing her as a frighteningly determined fanatic. The law was passed in a second vote. Arizona became the first state to ban smoking on buses and in lifts, museums, and public libraries. That was in 1973.

Other parts of the United States began to follow Carnes' lead. (It never entered anyone's head in Europe at the time that non-smokers should be protected from the second-hand toxins of their nicotine-addicted fellow citizens, or that smokers should be protected from the influence of the tobacco companies. Even today, Germany still lags far behind comparable countries when it comes to the fight against nicotine addiction.) Referendums were held in California, and by 1990 seven other states had followed suit. They failed in five of those states, although opinion polls had initially indicated that tobacco-control advocates were in the majority—until the cigarette manufacturers launched their massive campaigns.

Once again, the activists changed their approach. Rather than campaigning at a state level, they went local. They wanted what had been started by Carnes in remote Scottsdale to inspire similar campaigns everywhere. The aim was to make America smoke-free, city by city. There was little that the big companies could do in the face of such guerilla tactics.

At the same time, scientists were publishing new research that cast smoking in an even more negative light. It confirmed the fear that nicotine and tar damaged not only smokers themselves, but also those around them. This strengthened the activists' position. The often-repeated argument that smoking is an individual, personal choice that only impacts the smoker had been shown to be a dangerous misconception. One of the central arguments against the new rules thus collapsed.

The balance of benefit and loss when it came to smoking had shifted. Non-smokers were now not just fighting second-hand smoke merely as an annoyance, or out of concern for smokers' health. They were also fearful for their own wellbeing. Avoidance of loss is a stronger motivator than the prospect of a gain, which made the anti-smoking cause more persuasive.

When cities introduced restrictions on smoking in public, other municipalities followed suit. Visitors from neighbouring towns enjoyed the benefits of clean air, and demanded the same of their local mayors.[21] By 1990, more than 400 cities in America had introduced anti-smoking regulations, and the federal government gradually followed their lead.[22]

The power of the norm

More powerful than the direct effect of the restrictions was the message they sent out. The triumph of the cigarette was based on norms introduced into society by the propagandists of the tobacco industry. Afterwards, appearing with a cigarette was simply considered the done thing. As was striking up a conversation by offering someone a light, or blowing smoke rings to dispel boredom. Such habits had become so widespread over the years that no one ever questioned them. People often smoked simply because everybody was doing it, rather than because they really wanted to. The new rules changed those existing norms. Smoking a cigarette in the corridor while waiting for an appointment at a public office, or during university lectures, or while flirting with someone at a bar was suddenly no longer the norm, but a violation of the new prevailing order.

Of course, the new smoking bans were an encroachment on the freedom of smokers. Despite that, once they were introduced, they met with barely any resistance and a lot of approval. Even a majority of smokers supported the regulations—almost as if people had been waiting for a law to help them escape from the ubiquity of nicotine, which they were unable to do alone.

These views were borne out by research comparing towns with different levels of restrictions on smoking.

It found, for example, that more smokers quit the habit completely in localities where smoking was banned in restaurants than in those with bans that only covered publicly owned buildings. Even rebellious youths rarely saw the regulations as overly controlling. They saw them as a communal desire they wanted to share in: the stricter the restrictions in a town were, the more strongly both youths and adults agreed with the statement that second-hand cigarette smoke was annoying and dangerous. Once citizens got used to certain places being smoke-free, they appreciated the change and often demanded an expansion of the restrictions.[23]

The small initiative started by Betty Carnes with her *Thank you for not smoking here* signs had grown into a nationwide movement in just over a decade. Increasing numbers of people changed their habits and stopped smoking in public or quit completely. This reinforced the new norm. One city after another reached a tipping point — a term sociologists use to refer to the moment when so many people adopt a new behaviour that it becomes the expected norm throughout society.

Tipping points are difficult to achieve, as they require the entire population of a country or even a continent to be persuaded at once. But the smaller the scale, the easier it is to achieve a critical mass of people to lead the rest of the group by example. Once such a group has emerged, it serves as a model for others.

The secret to the success of Betty Carnes' movement was its modest beginnings. In a city such as Chicago or Berlin, or even in medium-sized towns, it would have been difficult to persuade enough people to show consideration for non-smokers. State laws failed precisely because they provided powerful industries with a bigger target to attack. But in the hidden corner of Scottsdale, Arizona, such change was possible. And once the ban on smoking at the hairdresser's, in bars, and at the library had been established in the little town, it spread from there to the surrounding cities, then to more distant ones, then to the whole of the state of Arizona, the whole of the United States, and eventually the entire world. More and more women and men kicked the nicotine habit, while fewer and fewer young people took it up in the first place. By the year 2000, the proportion of smokers among the American population had fallen from more than 45 per cent to less than 20 per cent.

A comparison with still-smoky Europe shows just how much Carnes and her co-campaigners had achieved. In Germany, where there were no local initiatives against tobacco use, the percentage of smokers in the population was almost double that in the United States in the year 2000. It was not a matter of money: when adjusted for purchasing power, the price of a pack of cigarettes was, in fact, higher in Germany than in the US. But the norms introduced by the big tobacco companies had not changed on the other side of the Atlantic.[24] Thirty-five per cent of adults in Germany were still addicted to nicotine, and far more Germans than Americans believed smoking to be unharmful. Pubs, meeting rooms, and railway carriages continued to be filled with clouds of smoke, and half the nation watched on in reverence as Helmut Schmidt, long after he was no longer chancellor, lit up cigarette after cigarette on television.

It took another seven years for Germany to pass any effective anti-smoking legislation. Almost every other country in Europe had learned by then from the successes in America and had enacted similar laws. The German Bundestag, the lower house of parliament, which has traditionally been under the influence of the powerful tobacco lobby, could no longer ignore the example set by France, Italy, and England.

German smokers were briefly outraged. But the fuss soon died down, and within a few weeks only one person still enjoyed the privilege of smoking whenever and wherever he wanted—the elderly Helmut Schmidt could not be denied his nicotine hit. Perhaps he felt wistful when he thought of the smoke-ridden Germany he had once governed, but, if so, he was the only one. No one else wanted to return to the old ways. The change had reached the last of the five phases that the philosopher Appiah believes every moral revolution must go through. It was now almost impossible for people to believe they had tolerated the old situation. Even more difficult to fathom was the reason why society had clung for so long to a practice that was clearly detrimental to everyone.

Despite the extremely addictive effects of nicotine, billions of people managed to beat their dependency, changing the societies they lived in in the process. They overcame not only their own biology, but also the power of a determined adversary: the tobacco industry. It seems that, when social norms change and individuals join forces, no obstacle is insurmountable. However, after progress has been achieved through struggle, it is never guaranteed to remain.

This is also borne out by the recent history of nicotine use. E-cigarettes and scented tobacco are the industry's latest ploys to encourage nicotine addiction in young people. Perfectly tailored to the tastes of teenagers, vapes are presented in bright colours and in flavours such as strawberry, mango, or even cheesecake. The seduction tactic works, and, since most e-cigarettes and vaping liquids contain more nicotine than ordinary cigarettes, they are more addictive, too. This has led to a renewed rise in e-cigarette smoking among 14-to-17-year-olds. The number almost doubled between 2021 and 2022 (admittedly, from a low base: from around 1 per cent to around 2 per cent).[25] If freedoms are not defended, they are stolen.

The great liberation

Ending a practice that has been recognised as harmful means opposing the power of lethargy and the fear of loss. However, even more difficult than cutting old apron strings is turning a new idea into reality. That takes more than just giving up old habits; it requires the courage to take a first step into the unknown.

It is often the simplest ideas that meet with the most resistance. A case in point is the idea that we should not exploit the Earth's resources to such an extent that life becomes impossible for future generations. Equally simple is the idea that every human is born with the same rights. Both statements roll easily off the tongue, and, of course, we recognise that they are true.

In the early 21st century, no one but the oldest diehards would claim that anybody has the right to elevate themselves above others due to their gender, skin colour, origins, or faith. But is everyone really treated as equally as we believe they should be? For centuries, people have struggled to get closer to the ideal of equality, but the question is: do they even want to get there? And the question of how to live sustainably currently divides societies.

That people will defend their privileges only goes part of the way to explaining these conflicts. Resistance is also born out of the idea that following the principles they agree with will mean people having to accept an unfamiliar and, for many, unimaginable way of living together.

In this chapter, I will present two historical struggles as examples of the kind of conflicts that regularly arise in times of upheaval. One ended in the abolition of slavery, and the other gave women their political rights. The struggles were based on the principle of equality, both began in the late 18th century, and they were connected to each other. Both movements were largely driven by women, and the activism of many female abolitionists led them to recognise how many rights were being denied to them and to consider how great a difference their vote could make, leading them subsequently to join the struggle for women's emancipation. In both cases, small groups first called attention to an injustice and demanded change, while the majority remained indifferent or hostile for a long time.

And yet those minorities managed to win over the world. People had kept slaves since the Stone Age, and societies almost everywhere were male-dominated.

Today, slavery is illegal in every country in the world, and, almost everywhere where elections are held, women vote and run for office as a matter of course. Powerful economic interests spoke against the abolition of slavery, and men were being asked to share power with women. How was resistance to these changes overcome?

Values, not knowledge

Knowledge can change societies. Without scientific research into the dangers of smoking, there would probably never have been a move away from tobacco. But knowledge is not always the crucial factor. In the struggle for the abolition of slavery, it was only important at the beginning, when Europeans began reading works such as the *Histoire des deux Indes* (*A History of the Two Indies*), which was published anonymously in Amsterdam in 1770 and republished 80 times, where they learned for the first time about the horrors of the trans-Atlantic slave trade. Fresh information was of little relevance in the struggle for women's political rights. The fact that women were oppressed by men and were hesitant to demand more power was not due to a lack of knowledge. Europe had seen enough successful queens to know that women could govern as skilfully as men.

The change was not driven by new knowledge, but

by new values. The upheaval began when individuals could no longer put up with the incongruity between what everyone knew to be true and what they saw. Under the influence of the Enlightenment and its call for people to use their powers of reason, the defence lines of cognitive dissonance were broken.

A typical testament to those beginnings was a pamphlet written after the French Revolution by the mathematician and philosopher Nicolas de Condorcet demanding equality of the sexes. There was no reason, he argued, to give priority to men, and there never had been. Condorcet confronted his contemporaries with their own inconsistency:

> Either no individual of the human species has any true rights, or all have the same ... Although liable to become mothers of families, and exposed to other passing indispositions, why may they not exercise rights of which it has never been proposed to deprive those persons who periodically suffer from gout, bronchitis, etc.?

Condorcet had used similar arguments ten years earlier, when he protested against the injustice of people of one skin colour enslaving those of another. However, he published that pamphlet under a pseudonym to avoid both censorship and hostility from the merchants based in France who were profiting handsomely from the slave trade. At the time, more than 10,000 African people a year were being shipped against their will to the French colonies in the Caribbean alone; the number of enslaved men and women working on the plantations of North and South America was in the millions.

A more courageous writer than Condorcet in that respect was Olympe de Gouges. That was the pen name of a butcher's daughter who was born in provincial southern France, and who had been married off early to a caterer against her will. When her husband died soon after, she changed her name and fled with their son to Paris, where she became a popular figure in fashionable Parisian society, educated herself through reading, and moved in literary and opposition circles.

She published her first play in 1784. It was set on a desert island where Zamore and Mirza, two former slaves and the title characters, had fled after Zamore had defended his lover from their owner when he had tried to rape her. The slaveowner had attacked Zamore with his sword, and Zamore had killed him in self-defence. The couple were caught, and Zamore was condemned to death. As soldiers were preparing the execution, the other slaves in the colony stood up to save Zamore's life. The result was a slaves' revolt.

Olympe de Gouges submitted her play to France's national theatre, the Comédie Française, and a committee unanimously authorised it for staging.[1] However, due to continuous delays, *Zamore et Mirza* never made it onto the stage at that time. It was seen as too innovative and, above all, as too scandalous.

Olympe de Gouges' demand that all human beings should enjoy the same rights from birth went against a millennia-old ideology. Hadn't even the Ancient Babylonians kept slaves, and didn't the Old Testament call explicitly for the enslavement of foreign peoples? Aristotle, one of the fathers of European philosophy, declared that those 'whose work is the use of the body, and ... this is the best that can come from them' were natural-born slaves. Aristotle and his teacher Plato both owned slaves. Even the great philosophers of the Enlightenment justified depriving people of their rights by violence. Thomas Hobbes and John Locke both lived in societies that enriched themselves by abducting people from Africa to the New World, and Immanuel Kant fabulated about a hierarchy of 'races', stating that Africans were particularly suited to a life of slavery. (Later, towards the end of his life, Kant changed his mind about this.)

It is no wonder that the Comédie Française preferred not to stage de Gouges' play. However, the theatre people had underestimated the writer's tenacity. She refused to give up. For years, she wrote letters, lobbied people, and sued the Comédie. In the revolutionary year of 1789, the theatre establishment finally relented, and *Zamore et Mirza, ou l'Esclavage des Noirs* (*Zamore and Mirza, or the Enslavement of the Blacks*), as the play had been renamed, was staged on 28 December of that year.

Some 2,000 people gathered in the auditorium of the Odeon Theatre in Paris to watch the play. The atmosphere was charged. Some of the audience applauded, and the *Moniteur* newspaper later praised the play as 'one of the most romantic productions ever to have been seen on stage'. But most of the audience saw Zamore as a criminal who deserved the harshest of punishments. The audience booed, whistled, and raged. Loud-voiced hecklers reminded the audience that it was a woman, of all things, who had dared to write this play.[2]

The hecklers had probably been sent to the premiere of *Zamore et Mirza* by the merchant families that earned well from the slave trade. It was common at the time for people to be paid to boo at theatre performances — trolling is not an invention of the internet age. But even if paid troublemakers were in the audience, why did the rest of the spectators join in with the booing? What was the explanation for the broad rejection of de Gouges' ideas? The Parisian theatregoers were not directly threatened by the demand that people of all skin colours be treated like humans. Very few of them would have ever met a Black person. And what was the harm in 'people without a beard on their chin', as one critic grumbled, writing plays? That was all beside the point. Olympe de Gouges did not attack people's privileges, but rather their prejudices. She held up a world to them in which it was not only Europeans who had reason, emotions, and morals, and showed them a future in which women had a voice that was also heard. Those who threaten old certainties risk being reviled.

Both these moral revolutions — one leading to the abolition of slavery, and the other resulting in the equality of women before the law — were still in the first of the five stages, when knowledge about an unacceptable situation spreads, but when most people are still able to deny that the problem exists. They remain trapped in a state of cognitive dissonance.

Olympe de Gouge's battle

'Men are born and remain free and equal in rights.' This is the opening proclamation of the *Declaration of the Rights of Man and of the Citizen* adopted by France's National Constituent Assembly immediately after the storming of the Bastille on 14 July 1789. It was not long, however, before plantation owners began to put pressure on the members of the assembly. It bowed to them, affirming that the declaration did not apply to the colonies, and certainly not to slaves.

It was also soon clear that the exclusively male members of the National Assembly really did mean *men*, and men only, when they drew up the declaration.

There was no call for the enfranchisement of women, or for women to be allowed to hold public office, pursue the profession of their choice, or own property in their own right. Even after the French Revolution, women remained subordinate to men.

On 5 October 1789, around 6,000 women marched on Versailles. They forced their way into the National Assembly and the Royal Palace. Many of them were traders in the markets of Paris who were angry at the ever-increasing price of food. When the king appeared and promised supplies of food, the mob insisted that Louis XVI accept the *Declaration of the Rights of Man and of the Citizen*, and that he and the government relocate to Paris. The king acceded to both demands.

In the following year, women's committees were set up across the country to demand civic rights for women. A certain Madame B. of Normandy wrote:

> There is talk of giving the Negroes their freedom; the people, almost slaves like them, will regain their rights. But what about women? The motto for women is: work, obey, say nothing … Unite, you girls and female citizens in the provinces, who are under the dominion of unjust and ridiculous customs … demand the abolition of the law that condemns you to poverty from the moment you are born.

Olympe de Gouges understood this anger well. She had experienced first-hand what it is like to be forced to live with a husband she had never wanted and to be dependent on him. She wrote another play, this time demanding the right of divorce for women. She repeatedly compared women's struggle with that for the liberation of enslaved people: 'Divorce allows the knots of marriage to be tied with flowers; without it they become the irons that the trembling slave gnaws at.' The Comédie rejected her manuscript.

In September 1791, she decided to tackle this evil at its root. She decided to make up for the National Assembly's omissions in its *Declaration of the Rights of Man and of the Citizen*, pointedly calling her manifesto the *Declaration of the Rights of Woman and of the Female Citizen*. Its structure and content followed that of the National Assembly's ambiguously titled document on the rights of 'man'. Each article of her text was formulated unambiguously, leaving no doubt as to which sex it referred to. Article one began, 'Woman is born free and remains equal to man in rights.' The declaration went on to say that with equal rights came equal obligations. Part of de Gouges' motivation was to clear this discrepancy up, once and for all. While Article 10 of the Rights of Man declared that 'No one may be disturbed on account of his opinions', she added, 'Woman has the right to mount the scaffold, so she should have the right equally to mount the rostrum.'

Olympe de Gouges called for the National Assembly to adopt her delaration. But the deputies had no intention of addressing her essay. Their reaction, if there was one at all, was derisive at best. But women's committees around the country began to adopt de Gouges' demands as their own.

More than a year later, on 30 October 1793, the National Assembly finally held a debate on women's civil rights. It was prompted by a motion to ban the radical Society of Revolutionary Republican Women for security reasons.

However, the speeches at the debate raised completely different questions. The deputy Jean-Pierre-André Amar asked whether women should be able to exercise political rights and 'meddle' in government affairs. As a left-wing Jacobin, he should have been politically close to the Revolutionary Republican Women, but his opposing view was patently clear: 'Do women have the moral and physical strength that the exercise of one and the other of these rights calls for? Universal opinion rejects this idea.' Amar then gave a prime example of ideological thinking based on opinions regardless of the facts, which was biased towards his own group: 'Man is strong, born with great energy, audacity, and courage … He alone seems to be equipped for profound and serious thinking that calls for great intellectual effort and long studies.' The role that was better suited to women was 'to prepare children's minds and hearts for public virtues ... such are their functions, after household chores'.

A deputy named Charlier then reminded the chamber that the mathematician and philosopher Condorcet had already refuted each of those points years earlier, so women could not possibly be denied the right of assembly. The chamber heard Charlier out, and then completely ignored his arguments. The assembly passed legislation that went far beyond just outlawing the Society of Revolutionary Republican Women. Its first article read: 'Clubs and popular societies of women, whatever name they are known under, are prohibited.'

Three days later, de Gouges was brought before the Revolutionary Tribunal. She was accused of having produced a poster calling for a plebiscite on a choice from among three potential forms of government: a unitary republic, a federalist government, or a constitutional monarchy. The tribunal found the 45-year-old defendant guilty of sedition. 'She wanted to be a statesman. It seems that the law has punished this conspirator for having forgotten the virtues that befit her sex.'[3]

'Woman has the right to mount the scaffold, so she should have the right equally to mount the rostrum,' Olympe de Gouges had written. On 3 November 1793 at seven o'clock in the evening, she was sent to the guillotine and beheaded.

The infectious nature of good ideas

Olympe de Gouges knew she was ahead of her time. 'If giving my sex an honourable and just consistency is considered to be at this time paradoxical on my part and an attempt at the impossible, I leave to future men the glory of dealing with this matter,' she wrote.[4] She would presumably have had similar feelings about her fight for the abolition of slavery.

However, the future was far closer than de Gouges may have thought. Ideas can spread like an infection — anyone who advocates an idea can pass it on to others.[5] That is why small groups, or even single individuals, can change entire societies in roundabout ways. The paths by which successful ideas take over the world are almost always so twisted that they can only be retraced in retrospect. They cannot be predicted ahead of time.[6]

Just one year after the publication of the *Declaration of the Rights of Woman and of the Female Citizen*, manifestos appeared in both Germany and England with similar content and even similar titles. Was that a coincidence? When Mary Wollstonecraft, the philosopher and mother of the creator of *Frankenstein*, Mary Shelley, published *A Vindication of the Rights of Women* in 1792, she was travelling regularly between London and revolutionary Paris. We do not know whether Wollstonecraft read the works of de Gouges, or if the two women ever met. But there is no doubt that Wollstonecraft was extremely familiar with the debates going on in revolutionary Paris.

We also do not know how much Theodor Gottlieb von Hippel, a Königsberg friend of Immanuel Kant, knew about de Gouges, but his book *On the Civic Improvement of Women*, printed in Berlin in the same year, 1792, certainly came to the same conclusion as she had—that women are not disadvantaged by their nature, but by society. And, like de Gouges, Hippel called for equal rights for women and men.

Mary Wollstonecraft's essay was a great success. It was praised in the newspapers of London and New York, American and French editions were printed, and novelists modelled their heroines on the liberated women whom Wollstonecraft described. The book became so popular that a biography published after her untimely death mentioned her in the title not by name, but as the author of *A Vindication of the Rights of Woman*.[7] Everyone knew who was meant without her name being referred to explicitly. Around the year 1800, there was much discussion of *A Vindication* and Hippel's *On the Civic Improvement of Women* in the salons of Berlin.[8] When women took up the struggle for their right to vote two generations later, they named Mary Wollstonecraft as their inspiration.[9]

The idea that all people should have the same rights regardless of the colour of their skin spread even faster. The pivotal figure was probably a man called Vincent

Ogé, a mixed-race member of the French colonial elite in the Caribbean. When he sailed from Haiti to Paris to pursue business claims in court in 1788, he was little noticed, except perhaps by the courts before which he brought his cases. A year later, the French Revolution began. As a person of colour, Ogé saw an opportunity to advocate for the rights of free-born people of African descent in the colonies. Another year later, Ogé returned to his native island. Once there, he spread the word about the debates in Paris over whether all human beings should have the same rights. Unhappy with the colonial government, which had not the slightest intention of abolishing white privileges, he took up arms. Ogé led a small uprising in 1791, which failed, and he was beheaded. At that, the enslaved people of Haiti rose up in revolt. It was precisely the situation that Olympe de Gouges had predicted in her scandalous drama, *Zamore et Mirza*.

The revolt swept across much of the French-ruled island within days. It was the largest slave uprising since Spartacus' ill-fated revolt against the Roman army. By contrast, the insurgents in what is now Haiti were able to hold their own.

In April 1792, the National Assembly awarded citizenship to all unenslaved inhabitants of the French colonies, regardless of skin colour. This meant that non-white males were given the rights that the National Assembly was still denying to all women. The decree failed to calm the situation in Haiti, as it was limited to those who were already free. Slavery persisted. The following year, the two commissioners who ruled the island as representatives of the government in Paris realised that radical change could no longer be avoided. Acting on their own authority, they proclaimed the abolition of slavery in Haiti, freeing half a million enslaved people. That was in August 1793. The first person of colour to join the National Assembly took his seat just one month later. He was Jean-Baptiste Belley, a freed slave (having bought his own freedom), who represented Haiti in Paris.

On 4 February 1794, the assembly followed the lead of its Caribbean commissioners: France banned slavery in all its colonies. It had been less than four years since Olympe de Gouges had shocked Parisians with her play *Zamore et Mirza*.

The prevailing opinion had completely changed in that short space of time. The moral revolution had reached the third of the five phases: no one could any longer ignore the contradiction between the stated aim of the young republic to make human rights universal around the world and the brutal oppression of humans on the plantations of its own colonies. At the assembly session when the law was introduced, a member called Delacroix exhorted the chamber not to 'dishonour itself by a long discussion'. The deputies agreed, and the law was passed by acclamation, with no vote needed.

The winding paths to change

The decision of 4 February 1794 marked the start of a new era. From the beginning of human history, people had always subjugated others, declared ownership over them, and found reasons that appeared to justify such a practice. Slavery was the norm, and that had been the case almost everywhere since time immemorial. Now, for the first time, a major country had declared its entire population to be free—and had moreover demanded that that right be extended to every person on the planet. The National Assembly replaced the norm of slavery with the new norm of equality and freedom.

However, the law was never enforced. The plantation owners refused to free their slaves, and the government in Paris lacked the means to impose the decree on its far-flung colonies. France had other problems to deal with. As we now know, fundamental change is never achieved with one single act; over a long period of time, every step forward is invariably followed by steps backwards. After the young Napoleon Bonaparte seized power in a coup in 1799, he reinstated slavery.

Napoleon prolonged the suffering of hundreds of thousands of enslaved people. Nevertheless, although he repealed the law in the French colonies, the world could no longer return to a time before the new norm of the universal right to self-determination. After further revolts, Haiti became the first country of formerly enslaved people to gain independence, in 1804. Paradoxically, Napoleon's reintroduction of slavery fuelled the growing worldwide condemnation of the practice.

To understand this in context, we can look at concurrent events in the British Empire, which was much larger than France's colonial possessions. In 1787, three years after Olympe de Gouges presented her *Zamore et Mirza*, a dozen activists met in a London bookshop to found the Society for Effecting the Abolition of the Slave Trade. Similar committees soon began to spring up in other British cities to denounce the cruelty of the slave trade.

The citizens of Great Britain had the right to petition parliament with their concerns. In 1788, the Manchester Abolition Committee initiated the first mass petition against the slave trade. Parliament more or less ignored it. That was no surprise—after all, the highly profitable sugar plantations were reliant on a constant supply of fresh labour. However, the breadth of popular support for the petition was unexpected. More than 10,000 people signed the Manchester petition—at a time when the entire population of the city barely exceeded 50,000.[10]

A book was then published, which, like de Gouges' *Zamore et Mirza*, told of oppression from the point of view of the oppressed. In the work, a former slave by the name of Oloudah Equiano recounted how he and his sister were taken as children from their village on the River Niger by man hunters, of the sea crossing surrounded by captives screaming with pain, and of plantation owners and slave traders who sold and resold him until he eventually managed to save up enough money from selling fruit to buy his own freedom. Equiano's autobiography gave a human face to the formally abstract demand for the legal equality of all people. Those who read it and identified with the narrator got a taste of what it was like to be the property of another person. And how could readers fail to identify with Equiano? After marrying a young Englishwoman, he led a perfectly normal British life in Westminster, London. Irrespective of his skin colour, the British thought of Equiano as one of their own. And they flocked to buy his memoir.

The committees organised another petition, this time a nationwide one. But the movement had now progressed beyond simply raising public awareness of the cruelty of the slave trade. It now targeted those who profited most from slavery. The campaigners called for a boycott of sugar from the Caribbean. Equiano himself distributed pamphlets calling for the first consumer boycott in history. The campaign explicitly appealed to women to join the boycott, as they usually managed the household budget. Women, who were often even more horrified than their husbands to learn of the destruction of families by the mass kidnappings, yet were not allowed to add their names to the petitions, now had a way to express their will.

More than 400,000 people across Britain joined the 1791 sugar boycott. The campaign achieved its aim of targeting the profits of the slave traders and their financial backers. Perhaps more importantly, it showed citizens that they did not have to accept a situation they saw as unjust. The moral revolution in Britain had now reached the fourth of its five phases: paths to possible action were now open.

Why was Britain able to progress so much further than France, mobilising a broad base of support for the anti-slavery cause on the northern side of the English Channel? In France, the issue remained largely confined to the revolutionary salons of the kind frequented by de Gouges. The general population did not have the capacity to even think about atrocities perpetrated on distant continents, let alone protest against them. During the years of famine as the decrepit *Ancien Régime* crumbled, and then in the chaos that followed the Revolution, people's priority was to ensure their own survival.

Firstly, the situation in Britain was very different. The country was politically stable, and industrialisation had already begun. People were migrating from the countryside to the cities, where they earned more, but had to endure miserable living conditions and long working hours, and were at the mercy of the factory owners. It was not difficult for workers in large industrial centres such as Manchester to imagine the slaves' plight. The fight for the human dignity of the oppressed labourers on the plantations reminded them of their struggle for their own dignity.[11]

Secondly, the abolitionists pursued a different strategy on each side of the Channel. While the French activists demanded an immediate end to the injustice of slavery, the British committees took a gradual approach. They first demanded that His Majesty's government put an end to the most obvious injustice — the abduction and transportation of people across the Atlantic. The British abolitionists abhorred slavery just as much as their French counterparts did, but they were willing temporarily to accept the situation of those already toiling in servitude on the plantations in order to provoke as little protest as possible.

The British tactics were cleverer, but their better success was also due in part to the uncompromising approach of the French. When the National Assembly abolished slavery, it set a moral standard against which every country subsequently had to measure itself. It did not matter that France failed to live up to its own standard. In fact, that gave the British abolitionists a further argument in favour of a slave-trade ban: it would allow Great Britain to show its moral superiority to the world.[12]

Another equally if not more important factor was that Britain was at war with Napoleonic France. This offered the opportunity to frame support for the anti-slavery cause as an act of patriotism. His Majesty's government could no longer afford to ignore such considerations. In March 1807, the situation came to a head. At the height of the Napoleonic Wars, the Abolition of Slavery Act was passed with an overwhelming majority in parliament. The British Navy began treating slave traders as pirates, and a special squadron was set up to patrol the Atlantic transportation routes. Over the following decades, it freed more than 150,000 Africans who were being smuggled to the New World below deck. And, since Britannia really did rule the waves at that time, the slave trade gradually dried up as a result.

It was a victory for a moral principle over economic interests. Britain owed a significant portion of its wealth to the labour of enslaved people in the Caribbean, and by the early 19th century, both the sugar and slave trades had become more lucrative than ever before. In 1805, a good 20 per cent of British trade consisted of business from its plantations.[13] That is almost exactly the same as the automobile industry's current share of Germany's foreign-trade volume.

The women rise up
Slave trading was made illegal throughout the British Empire, but there were still more than 700,000

enslaved people working on the plantations. This was changed by the sugar boycotts that started in 1824. This time, the protest was led by women. They set up committees across the country to coordinate their campaign. Grocers began to advertise that they sold sugar produced without slave labour. Sugar bowls bearing the gilded motto 'East Indian sugar not made by slaves' became fashionable in the parlours of the middle classes. And sugar bowls with illustrations of dark-skinned people in chains appeared on the kitchen tables of the working class.

The plantation owners were not going to give up easily, however. They mobilised a powerful lobby group that they called the 'West India Interest'. It united those who profited from slavery in the Caribbean—merchants, bankers, politicians, and the plantation owners themselves. Its aim was to prevent the change for as long as possible. 'The Interest', as the group became known, used the same strategy and methods as fossil-dependent industries do today.

Parliament passed unbinding resolutions recognising the difficult situation of plantation slave labourers and promising betterment. Committees were formed—with representatives of 'The Interest' making up the majority of their members. Media campaigns were launched, journalists were courted to persuade them to write of the benefits of slavery, and opponents were defamed as ideologues or rabble-rousers. One cartoon from 1826 shows happy slaves dancing on a distant palm-covered island while, on the other side of the ocean in England, poor children are shown starving and freezing to death. A man is peering through a telescope, trying to see what the slaves' lives are like, but a dishevelled pastor is holding a scene of horror in front of the lens.[14]

However, the women activists were able to counter the propaganda of 'The Interest'. No one could deny the reality of conditions on the plantations any longer. The issue dominated the public debate. More and more citizens signed petitions calling for the immediate emancipation of all enslaved people. And the slaves in the colonies were aware of the support they were receiving. In 1831, 60,000 enslaved people went on strike in Jamaica. The rebellion was brutally quashed, which only served to add more momentum to the anti-slavery campaigns back in the mother country.

In 1833 alone, activists collected 1.5 million signatures, and handed the petitions to parliament in London. The country had never seen a citizens' movement of such size. One petition alone bore the signatures of 350,000 women. The women sewed and pasted the name-lists together into a huge document that was half a mile long when unravelled.[15]

Parliament gave way in the same year. It declared the abolition of slavery in all British colonies. After a transitional period, 800,000 enslaved people were emancipated, and huge compensation payments were made to their former owners. Half a century after the French Revolution, the goal set out in the National Assembly's resolution of 1794 had finally been achieved. The fundamental right to self-determination had gained acceptance around the globe, and slavery was outlawed on all five continents. Wherever the Union Jack was flown, no one could legally claim ownership over another human being. The anti-slavery campaign and the sugar boycott had shown the power of determined citizens in the face of even the greatest resistance.

A domino effect followed: each change triggered another. Knock-on effects often appear in seemingly unconnected and unexpected areas, and the same was true of the fight for the rights of the enslaved people on the plantations begun by a tiny group of determined activists. In campaigning for the rights of enslaved people, women realised the power they held. They learned they could play a part in public life and started demanding their right 'to mount the rostrum', as Olympe de Gouges put it. They realised that the principle of equality must also apply to them. If it was the tradition for women to stay in the background, then too bad for tradition. The emancipation of the slaves made the emancipation of women possible.

An instruction manual for a better world

When trying to do the right thing in spite of one's own laziness, Germans speak of overcoming their 'inner pig-dog'. That colourful, if slightly vulgar, phrase sums up the way we make change unnecessarily difficult for ourselves. And it has an interesting history. People could be forgiven for thinking that 'pig-dog', or *Schweinehund* in German, originally referred to a mythical monster with the legs of a dog and the head of a pig. But the *Schweinehund* was a real animal. The word used to refer to aggressive hunting breeds, such as bulldogs, which were used to hunt wild boar. Its meaning was later extended to include malicious people.

The first recorded use of the phrase 'inner pig-dog' was at parliamentary session at the Reichstag in Berlin. On 23 February 1932, during a heated debate, the young Social Democrat Kurt Schumacher denounced Nazism as 'a continuous appeal to the inner *Schweinehund* in human beings'.[1] The president of the

Reichstag called him to order, but Schumacher was undeterred, going on to say that Nazism was unique in German history for its success in 'ceaselessly mobilising human stupidity'. Schumacher's understanding of the phrase 'inner *Schweinehund*' was that it was a force lurking inside every human being that is harmful to the self and is unleashed by lies and hate speech.

A few weeks later, the defence minister of the *Reich* picked up the phrase. But he had a different interpretation of the 'inner *Schweinehund*'. In a radio address, General von Schleicher praised young soldiers who had 'demanded something extraordinary of their bodies ... and completely conquered their inner *Schweinehund*.' So, for him, the inner *Schweinehund* was merely a weakness of will. If you are ruthless with yourself, you can overcome your indolence. Nine months later, Hitler seized power. The message that the German people were to hear repeated thousands of times was that the will must triumph, and that victory would not be achieved through reason, but through toughness — thus 'overcoming their inner *Schweinehund*' became a recognised virtue of German soldiers.

That is the way the phrase is still used in contemporary German. What has changed is what we now consider reprehensible or undesirable behaviour. No one today believes it is their duty to fight to the last breath for their nation and its leader. Today, we blame our inner *Schweinehund* if we spend the day on the sofa eating chips instead of going out for a run, despite our best intentions. And we also see it as the culprit when people waste the planet's resources out of laziness or greed, or give free rein to their anger on social media.

What has not changed is the methods we use to try to change ourselves and the rest of the world. We believe it is enough to recognise a problem and then make a firm resolution to remove it. So we decide—usually in the New Year—to finally cut down on drinking, do more exercise, and stop putting off necessary jobs. Just like in General von Schleicher's time, we declare war on our inner *Schweinehund*—a war that can only be won with sacrifice, self-denial, and a virtually superhuman act of will, if it can be won at all. No wonder, then, that we are happy to give up most such agonising battles as soon as the first temptation comes along.

We allow that other people might simply be ignorant of the truth—and see it as an act of benevolence to put them right. Once we have kindly provided them with the information they lack, we believe they have a moral duty to act accordingly. And if they don't? We are disapproving of people who persist in their habits although they know better and could, in our opinion, change their behaviour. We consider them weak-willed, selfish, or both—and in any case incapable of reining in their inner *Schweinehund*.

A culture of change

However, mere willpower can do little against human indolence, and toughness can do nothing. General von Schleicher's fantasies about the soldiers, which we still adhere to, are far removed from reality. If there is such a thing as an inner *Schweinehund*, it cannot be defeated by brutality. It takes cunning and intelligence.

The longer people have a habit, the less conscious of it they are. Habits are controlled by triggers, not intentions. Reluctance to change them is not human weakness, but the result of the way our brains are organised.

Millions of years of evolution cannot easily be erased, so commands to suddenly start acting differently usually have one result — resistance. Inevitably, the fronts then harden. Resistance is often born more out of a feeling of helplessness than out of conviction. That feeling is highly dangerous, since it can be exploited by demagogues to stir up hate and destructiveness. That is what Kurt Schumacher was referring to when he accused the Nazis of unleashing the inner *Schweinehund*.

Where, then, can we find the solution? Very few people are born with the ability to adapt to change quickly. Nature designed us for life in a stable environment, and that is what our brains are programmed for. But agriculture, housebuilding, city life, the creation of nation-states, differential calculus, and music — none of that is in our genes. Culture is what enables us to achieve all that. Culture encompasses more than just tangible achievements. Above all, it is a toolbox for the mind. It is a constantly evolving system of ideas, methods, and values enabling us to survive and innovate in an increasingly complex world.[2] Culture expands our thinking, emotions, and decisions, just as a hammer grants superhuman strength to a human hand.

So culture can also free us from our impotence in the face of ever-faster change. We need the mental tools to help us recognise when old convictions conflict with reality, to help us in giving up old habits, and to give us the courage to strike new paths. We need a culture of change.

According to the soldierly mentality described above—which is still widespread today—individuals must vanquish their inner *Schweinehund* alone. A culture of change says otherwise. It recognises that profound changes can only be achieved by working together. Most attempts to change behaviours or attitudes fail because individuals expect too much of themselves by attempting to shoulder the burden alone. Just as no one can build a skyscraper by themselves, we need each other if we are to give our lives a new direction.

Four steps to new beginnings

A culture of change must take into consideration how our minds react to the new. The previous chapters described various factors that make change easier or more difficult. A culture of change takes those factors seriously and learns the lessons of history. The revolutions that societies go through do not repeat themselves in the same form, but they do repeat the same patterns of change. We can take encouragement from those historical experiences because they are proof of human beings' adaptability.

There are four crucial steps to change. The first enables people to recognise their own long-term interests. As a result, they no longer feel like a football kicked around by outside forces, and are in control of their own actions. The second step has to do with habits. It deals with replacing old routines with more beneficial ones. In the third step, new attitudes and behaviours are reinforced as they spread through society. The fourth and final step broadens the scope of the change. It is about our expectations and the stories we tell ourselves about the future. All four steps have a common goal: people should not experience change passively, but actively shape it. The four steps are described in more detail below. Practical examples show how the principles can be implemented. They are not presented as the only possible solution, but as suggestions.

First step: Better decision-making

A culture of change must be based on freedom of choice. No one must be forced into a change because others believe it is good for them. Nor should they be converted to a cause. A culture of change has the courage to believe in the power of reason.

It recognises that people often act irrationally and against their own best interests. In the past, when people made nonsensical decisions, it could be explained by a lack of knowledge. In a culture of change, by contrast, mistakes are not put down to ignorance, but to the illusions created in our minds, even when there is a wealth of real information available—or perhaps precisely because of that. Our internet browsers spit out detailed and usually well-presented information on almost any conceivable question. However, as described earlier, we stumble at the interface between facts and knowledge.

One pressing challenge, then, is to learn to handle information better. Today, we receive most of it online, and that is precisely the area in which research has revealed shocking shortfalls, for example in Germany.[3] A comparison of the digital literacy of eighth-graders across 34 countries revealed that 40 per cent of young people in Germany 'can barely do more than click on a link, type a WhatsApp message, or scroll through Instagram and TikTok', as the national coordinator of the study put it.[4] The digital literacy levels of German school students actually fell between 2018 and 2023. Most are now incapable of recognising fake news or scam emails, and are unable to find the information necessary for a given task even on a trustworthy website. (The figures were even worse for the United States, where the decline in digital literacy levels was even greater than in Germany.)

In Finland, by contrast, children are taught in elementary school how to recognise false information on the web. In secondary school, they have lessons in critical thinking and are taught to understand statistics correctly.[5] The students are not taught what is true and what is false: the lessons equip them with the tools to recognise the difference themselves. In this way, members of the next generation are prepared to defend themselves against manipulation. The success speaks for itself. Finland not only has one of the best education systems in the world, but it also tops international league tables when it comes to adult digital literacy.[6]

Such measures are an encouraging start. But the internet intensifies our tendency to delude ourselves. The real causes lie much deeper. More important than the ability to spot manipulation by others is the skill of recognising the traps our own minds set for us.

Realistic and responsible decision-making requires clear thinking.[7] Twentieth-century psychologists were doubtful that thinking could be taught; they believed people were only able to learn to solve specific tasks.[8] That would mean that the mind can be clouded by prejudices and fallacies every time we encounter a new problem. However, recent research has yielded more optimistic results.[9] They show that it is indeed possible to learn to use our minds better in general.[10] People make better decisions when they are aware that they might fall victim to a certain cognitive bias.[11]

Our education system ignores all these facts. It is still based on standardised knowledge. However, in this age of information overload, memorising facts is of secondary significance. Understanding how the mind processes information is far more important. Only those who are aware of the enticements and deceptions lurking in the flood of information and rumours we face today can act sensibly. Metacognition—awareness and understanding of one's own thought processes—is the central skill of our age. But it is never taught.

Metacognition can be learned. People of any age can grasp the meaning of statistics if they are explained using everyday examples and the data is presented clearly. They learn to think in terms of possibilities, which is essential if we are to interpret issues correctly in our complex world.[12] Avoiding fallacies can also be improved by practice. Several studies give cause for optimism. Subjects in one experiment played a specially programmed video game in which the aim was to find a missing person who was involved in a crime.[13] To solve the case, participants had to think impartially. After only an hour of playing the game, subjects were better at solving management problems, even though they were structured very differently from the on-screen detective game. And that desirable effect was lasting.

Three months after the experiment, the subjects still displayed above-average skills in strategic decision-making.

Preschools and schools, universities, and in-company training, even online courses, can teach people how the mind reaches its judgements — and provide people with the tools to make better decisions. Such a programme of education is complex and expensive, but its effects are long-lasting. There is no alternative if we are to overcome our self-destructive tendencies. Our society can only continue to exist in the future if we invest far more in people than we currently do.

Second step: Changing habits

There can be no will to change without understanding. But understanding alone is not enough. No one in their right mind believes it is possible to learn gymnastics or improve their rowing stroke through understanding and willpower. That can only be done with patient practice and training. Appeals are also ineffective in getting people to change patterns of behaviour that have been ingrained for years. Irrespective of whether the appeals are made to people's reason, conscience, or their own interests, research shows that even the best-planned campaigns to persuade people to save electricity, eat less meat, or use their cars less for the sake of both the planet and their own budgets have almost no impact.[14] Most of those interviewed reported having attempted to change their behaviour, but having soon relapsed into their old routines. The difficulty is not in adopting new habits, but in casting off old ones.

An athlete can break a stubborn, undesirable pattern of movement with the necessary support, such as regular feedback and encouragement from a coach. At first, she may feel awkward, and her coach will have to remind her repeatedly of every single muscle movement to keep the unwanted routine from returning. With time, her own brain will take over the job of the coach, and the new movement will become more fluid, require less concentration, and eventually become second nature to her. She is then over the hill and has progressed from an undesired stable state to a new, better stable state.

Changing habits is a similar process. Encouragement and feedback do not necessarily have to come from a person. In one example from earlier in this book, a light went on to remind the staff in an intensive care unit to wash their hands before every interaction with a patient. The crucial point was that the display did not show failures, but successes. Hand hygiene gradually became second nature to the doctors and nurses. Similarly, displays showing residents' successes in saving electricity in their building have also been shown to be effective.[15]

The concept of the carbon handprint is also based on the principle of positive feedback. This way of measuring environmental impact was pioneered mostly by Nordic companies.[16] The more famous carbon footprint is a measure of the greenhouse gas emissions generated by a certain behaviour. The carbon handprint quantifies climate change mitigation potential. It does not give carte blanche to anyone; it provides an incentive by indicating how much more can be gained from further improvements.

The handprint would be easy to introduce to the general public. For example, an indicator of the environmental impact of every type of packaged food, from camembert to vegan bratwurst, could be printed next to the nutritional information.[17] Works canteens could put up signs prominently reminding those workers who join the line for the pasta dish that they are impacting the environment with half a hundredweight of carbon dioxide less than their colleagues waiting for the beefsteak.[18]

A similar method in the field of behavioural economics is known as nudging. It is a way of getting people to take a certain course of action by making the relevant decisions easier. For example, the button to click for a certain operation on a website might have a more eye-catching colour than the alternatives. However, the nudging effect is very short-lived. It works by influencing people anew before every decision. The suggestions that I give here, while they are meant to offer help on a temporary basis, are aimed at changing habits permanently. Unlike nudging, whose effect is often unconscious, they require both knowledge and a will to change.

Routines are activated by triggers, not intentions. A fresh context often helps to break them down. Supplying communal breakfasts in French schools, at which young people enjoyed healthy food in good company, rather than eating junk food in front of a screen, led to a sustained 50 per cent reduction in the proportion of children who were overweight. A simple browser extension such as the one described in a previous chapter, which delays the loading of certain websites by a couple of seconds, gives control over their own online behaviour back to users.

Even the most stubborn of habits can be changed with patience and realistic expectations. However, anyone hoping to solve the problems of the world more quickly with appeals to morality and calls to conquer an inner pig-dog is inevitably bound to fail.

Third step: Start chain reactions
Changes happen suddenly. Inertial forces, feedback effects, instabilities, and tipping points bring about sudden changes of state. People infect each other with their behaviour and attitudes, which makes unexpected transitions possible.

After the dangers of smoking were proven in the late nineteen-forties, nothing happened for four decades. As described earlier, cigarette consumption actually increased, despite the emergence of ever-more shocking facts about the harm caused by nicotine and tar. But when the change did come, it came very quickly. Between 1985 and 1990, cigarettes largely disappeared from public life in the United States, and the same occurred in Europe between 2005 and 2010. An avalanche had been set in motion.

Avalanches are a type of chain reaction. When enough snow has accumulated and a small amount begins to slide, it takes the neighbouring snow with it. That, in turn, drags more of the snowpack along, and so on until the entire slope is set in motion—the system suddenly changes its behaviour. Avalanches are almost impossible to predict. There are many factors—in this case, the slope's gradient, the temperature, wind, composure of the snowpack, and even the shape of the snow crystals—that determine when and if a chain reaction begins and gathers momentum.

Societal change behaves the same way. But it is even more unpredictable, since human beings are infinitely more complex than snow crystals. Overly simple recipes for certain and rapid change in an organisation, or an entire country, are therefore rarely worth our attention. We are not able to predict how attitudes change. When researchers held a 'forecasting tournament' among 100 teams of leading social scientists, who were asked to predict how the views of Americans on gender equality, racism, life satisfaction, political polarisation, and ideological preferences would change over the period of one year, the results were sobering. The social scientists' predictions were barely more accurate than if they had just rolled a dice.[19]

The dynamics of chain reactions lead us to three important conclusions. Firstly, the similarity between societal change and physical chain reactions shows us that we must have patience — and that we can have hope. Just because change *seems* impossible, it doesn't mean it actually is. Snowpacks take time to accumulate. Before they can plummet down the mountainside in the form of an avalanche, they not only have to reach a critical mass, but the structure of the ice crystals also has to change. Those processes take place deep in hidden layers of the snow. Without them, the later chain reaction would not be possible.

The British environmental activist and journalist Georges Monbiot wrote that the experience of all effective movements shows that their success lies in preparing for the moment of change.[20] It can come unexpectedly. Sometimes, the role of an entire generation is simply to prepare, develop arguments, tell stories, and organise campaigns for the moment when their successors will drive real change.

Secondly, the dynamics of chain reactions enables us to analyse the phenomena that are critical for societal change. Just as we can estimate the critical mass necessary to trigger an avalanche, we can consider what proportion of the population have to change their habits or attitudes for a new pattern of behaviour to permeate throughout society. The threshold is surprisingly low. Empirical studies and theoretical simulations show that only 17 to 30 per cent of the members of a group can change the minds of the majority and bring about lasting change.[21] For example, if the proportion of women working at a certain position in the hierarchy of an organisation reaches that level, people begin to see it as normal for women to occupy such jobs.

It might be argued that even winning over a fifth of the population to embrace radical change is an unrealistic goal. For example, to persuade the population of Germany to accept a more resource-friendly diet as the norm would mean convincing 20 million people to eat little or no meat. That seems like an impossible ask.

We have already learned that there is a way out of this dilemma. Avalanches rarely start big. They almost always begin with a small snowslide. When the snow starts to move, it drags masses from elsewhere with it. The ubiquitous presence of cigarettes was initially curbed only on the streets of Scottsdale, Arizona. Other insignificant places gradually followed the lead and introduced such a policy, then big cities, then entire states, and eventually the whole country followed. The smaller a community is, the easier it is to reach the critical threshold and then serve as a model for other groups.

Other processes of change follow the same dynamic. Swedish and American studies have shown that homeowners are more likely to invest in solar power modules and efficient heating systems such as heat pumps if their neighbours already use them. When the new technology has spread to the whole street in this way, it can go on to spread to the adjacent neighbourhood.[22]

All too often, activists try to persuade the whole world — or at least entire countries — to change. A more realistic strategy is to start by creating islands of change. Unfortunately, new norms can often, or sometimes only, take hold when a population is divided into smaller communities.[23] Once individual groups start to adopt the new thinking or behaviour, the likelihood increases that more and more groups will be caught up in a chain reaction.

Thirdly, the dynamic of chain reactions refutes the idea that individuals cannot achieve anything on their own. Olympe de Gouges helped women to gain their civil rights and enslaved people to gain their freedom by infecting others with her ideas. Betty Carnes freed the world from tobacco smoke by acting as a model for others to emulate. A more prosaic example is the idea that there is no point in casting a ballot in elections because one vote among many millions can make no difference. This sceptical attitude fails to take into consideration the fact that every individual's behaviour can be multiplied through contagion effects. As the American sociologists James Fowler and Nicholas Christakis proved, single individuals can influence up to 100 others with their voting decision, and their influence goes beyond just family and friends, reaching friends of friends in a domino effect.[24] The effect is even stronger if a group acts as a unified role model.

The dynamic of chain reactions thus provides us with an unexpected solution to the dilemma described in the chapter on illusions of freedom: Should I avoid unnecessary air journeys for the sake of the climate, although the amount of pollutants emitted by that one flight is negligible? The answer is yes. And especially so when others are merrily boarding their flights, because a decision to fly or not influences others; it shifts the norms in society. And whether we destroy the foundations of our life on Earth or preserve them depends solely on the norms that shape our behaviour.

Fourth step: Tell good stories

> 'Creating a ship is not a case of weaving the sails or forging the nails or reading the stars, but of giving a taste of the sea.'
> **Antoine de Saint-Exupéry**[25]

The British author Julian Barnes built two contrasting biographies around the documented facts of the life of his fellow novelist Gustave Flaubert. In the first version, Flaubert is a luminary of French literature and the darling of Paris society. In the second version, he is portrayed as an artistic failure and a human wreck.[26]

Humans love stories. Furthermore, humans live within stories. We need stories to guide us. When we want to understand a set of facts, we construct a plausible story to explain them. We ask whether our experience corresponds with what we know about the world. Is it coherent? Is it developing as we expect a good story to unfold? Facts inform us, but they only gain meaning when they are arranged into a story. And only then can they move us to action. The problem is that the same set of facts can be woven into very different stories.

Even nuances in the way things are represented influence the way we register facts — and thus also the way we behave. There is no doubt that regular exercise prevents illness, but it makes a big difference whether saying so is framed as a reminder that exercise is good for your health or as a warning that lack of exercise increases the risk of obesity, diabetes, and heart attacks. We process possible gains and possible losses differently. Our brains associate proactive behaviour with rewards, and evasion and flight as threat avoidance. Those reactions are rooted deep in our midbrains, and take place at the subconscious level.

On the one hand, fear provokes more attention than the hope of a reward. On the other hand, fear paralyses. It stops people from engaging in certain behaviours, but does not motivate people to show initiative or to change their habits. If the message includes the hope of better health, it will persuade more people to take up regular exercise than if it is expressed as a warning.[27]

The stories we currently tell ourselves about the future are fear-driven horror stories. Wars, environmental disasters, economic collapse, and the end of democracy—our television screens feed us a daily diet of images of the apocalypse, and online media update us constantly with terrible news. It is hard to escape their pull.

Activists can also be guilty of scaremongering. For example, their warnings about the climate crisis often go as follows: if we do not stop burning oil and coal immediately, the world will end in a ball of fire. Deserts will spread, billions of people will starve, the polar icecaps will melt, and the oceans will flood entire countries. There is only one way to prevent the collapse of civilisation at this late stage—we must all make sacrifices. We will no longer be able to enjoy the comforts and security of a prosperous society. Belt-tightening is painful. But there is no alternative if we are to save ourselves and the world.

That message attracted some attention for a while. But it had little effect. Fear feeds anger and paralysis without offering any perspective for the future.[28] It does not motivate us to change habits and thinking. The more often we face fear, the more numbed we become to it. Fear evolved as an automatic response to short-lived, concrete threats such as those from predatory animals.[29] The threats we are now facing are long-term and barely tangible. Nature did not wire our brains to deal with such threats.

The following narrative also tells of the challenge we will face due to the climate change: oil, gas, and coal have served us well. In the rich countries of the world, they helped us achieve a level of prosperity previously unknown in history. But now it is the time for us to part with them. Oil and gas have made us dependent on supplier countries, and therefore vulnerable to blackmail. We want to live in freedom and security. We also want to leave a liveable world behind for our children. And we have found safe—and cheaper—alternatives: solar and wind power.

Transitioning will be expensive, but we can shoulder that financial burden if it is distributed fairly. We have already seen that further increasing prosperity in rich countries does not make people any happier.[30] Therefore, the necessary transition to an economy that is less damaging to the planet also offers an opportunity. Halting the constant increase in consumption, or even reducing it, will not only lower the amount of resources consumed, but also offer the freedom to engage in that which really improves our quality of life.[31] When people switch to a more environmentally friendly lifestyle, they also benefit themselves.[32]

This depiction of our future world is less dramatic than the bleak scenarios described by climate activists. But it is just as well founded and just as true. Most importantly, it is immeasurably more encouraging. It does not drive people into resignation, but moves them to act. In the same way, other social upheavals such as the advent of artificial intelligence and the drastic ageing of society can be depicted either as doom scenarios or as calls for a new beginning.

Which perspective we take is crucial, because the stance we take today will determine whether we have to passively suffer our future or can actively shape it. Fear causes paralysis. Those who know clearly what they want can strive for their goal. A culture of change does not mean sugar-coating the world. It names the dangers. But those threats are not the end of the story; they are the starting point for narratives that eventually lead to the question: what do we want our lives to be like?

The nature of hope

A shocking number of people in industrialised nations have given up hope. They see their societies divided by an ever-growing gap between the poor and the unfathomably rich. They see new digital channels spreading division and bare-faced lies, and they see increasing numbers of women, men, and children fleeing war, environmental disasters, and the gradual destruction of their livelihoods. They no longer believe that their children will inherit a liveable world. Many have succumbed to resignation. Others seek refuge in ideologies, or ally themselves with demagogues who promise easy solutions and a return to the supposedly good old days.

However, people who feel discouraged in this way fail to understand the nature of hope. Hope is not the blind expectation that everything will turn out fine in the end. Hope is a recognition that the future is not yet written. Hope means having understood, firstly, that predictions can only look so far into the future, and, secondly, that we can influence what lies ahead. Hope means believing that the world is unfinished.

There has never been a time when it was less appropriate to consider the future as written in stone. Never have so many people had so many opportunities, and humanity has never had such a wide body of knowledge at its disposal. Technology is developing at breakneck speed, and it gives all of us powers that past cultures would only have ascribed to the gods.

We are living in unusual times. The distinguishing feature of the conditions we live in now is that they are so unstable. Instability creates unpredictability—but it does not necessarily mean that things will get worse. They could just as easily end up being better than we expect. We simply do not know. So, from both a moral and an intellectual point of view, pessimism is not an option.

'Hope is ... not the conviction that something will turn out well, but the certainty that something makes sense, regardless of how it turns out.'[33] When the Czech dramatist Václav Havel said this in an interview in 1986, he had spent three long terms in prison because of his intransigence towards the regime in Prague. No one suspected at the time that a peaceful revolution would sweep the communists from power in just three years' time, and that Havel would become the first postwar president of a free Czechoslovakia.

In this book we have seen what we need to do to bring about change. It is crucial that we act together. People adopt new habits and new convictions when they support and encourage each other to live according to their moral values. A culture of change relies on motivation and information. It bridges the gap between those who lead and those who follow. In such a culture, people no longer need to feel ashamed of their habits and mistakes, and they no longer need to feel afraid about the future. A culture of change allows us to experience the feeling of achieving success, initally by taking small steps and making small gains — success that comes from daring to try new things. It taps into our justified pride at being the most cooperative, intelligent, resourceful, and adaptable creatures on the planet.

We are in danger. But humanity currently still has all the means necessary to master even existential challenges such as climate change or the rise of artificial intelligence. What we have lacked for so long is a sense of reality. We have been avoiding change. So isn't now the time for us to do all we can to open ourselves up to new things?

Notes

On top of the volcano
1 The same phenomenon was observed following the Chernobyl and Fukushima nuclear disasters, and, to a lesser extent, after the devastating floods along the River Ahr in western Germany in July 2021. There is even a local word for those who refused to follow guidance and returned to the Chernobyl death zone to settle there: *Samosely*, meaning literally 'self-settlers'.
2 Aiken and Keller 2009.
3 For example, 8.7 million premature deaths in 2018. Vohra et al. 2021.
4 Hertig et al. 2023.

A paralysed society
1 Stephens 1854.
2 Medina-Elizalde and Rohling 2012; Kennett et al. 2012; Evans et al. 2018.
3 Initial, still-imprecise indications of a period of climate change had already been published: Hodell, Curtis, and Brenner 1995; Gill 2000; Haug et al. 2003.
4 Weiss 2017; Cook et al. 2022.
5 Evans et al. 2018; Haug et al. 2003.
6 Cook et al. 2012; 2022.
7 McIntosh 2007; Farmer, Sproat, and Witzel 2004; Mithen and Black 2011.

8 Tainter 1988.
9 Maier 1968.
10 Tainter 1988.
11 economist.com/finance-and-economics/2023/08/17/the-german-economy-from-european-leader-to-laggard.
12 Vogel and Hickel 2023.
13 futureoflife.org/open-letter/pause-giant-ai-experiments; see also Russell 2020.
14 Strauss 2023.
15 Acemoglu 2021.
16 Orlowski-Yang 2020.
17 The actual ratio is, in fact, even worse, since not every person between the ages of 20 and 67 is in paid work.
18 Kaltenberg, Jaffe, and Lachman 2023.
19 European Commission 2024.
20 statista.com/statistics/589607/average-prices-new-cars-germany.
21 Even with a relatively moderate annual mileage of 15,000 kilometres, the higher purchase prices of electric cars are amortised within less than five years. 'Kostenvergleich: E-Auto, Benziner oder Diesel?' 2023.
22 This will be less surprising to those familiar with a different experiment, which was carried out a little earlier in the same year, 2023. Fresh data published by the *Bundesumweltamt* (German Federal Environment Agency) showed that introducing a speed limit on the autobahn would save three times as many carbon dioxide emissions as previously assumed. Survey interviewers gave this information to respondents who had been selected as representative of the German population. The information had a minimal impact on their opinions. The only group that became more open to the introduction of a speed limit were that minority

of respondents who had already expressed a willingness to change their behaviour in other areas of life in order to protect the climate. See also projekte.uni-erfurt.de/pace/topic/special/70-tempo.
23 Dechezleprêtre et al. 2022.
24 Lorenz-Spreen et al. 2023.
25 Przybylski and Weinstein 2017; Twenge and Campbell 2018; Twenge 2019; Pandya and Lodha 2021; Haidt 2024.
26 Engels and Grunewald 2017.
27 bitkom.org/sites/default/files/file/import/Bitkom-Charts-PK-Datenschutz-22092015-final.pdf.
28 Acquisti and Gross 2006.
29 Nolt 2011. For a more recent estimate, see Carleton et al. 2022. It might be argued here that companies are responsible for far more damage than our own lifestyle. However, that argument also falls down because any industrial emissions can be attributed to the goods produced. Everything industry produces ultimately benefits one or more consumers. Anyone who eats meat cannot claim to have nothing to do with the death of the animal they're eating.

First Illusion: We are realists

1 Festinger, Riecken, and Schachter 1956.
2 FitzGerald, Dolan, and Friston 2014; Friston 2018; Clark 2019; Yon, Heyes, and Press 2020.
3 To find this, type 'rotating mask illusion' into your internet search engine.
4 Wittgenstein 2009.
5 It adorns the 12th-century Airavatesvara Temple in Kumbakonam.

Second Illusion: We love novelty

1. Zajonc himself credited the work *Vorschule der Ästhetik* (*Introduction to Aesthetics*), published in Leipzig in 1876. Its author, the psychophysicist Gustav Fechner, had considered why people from different cultures have widely differing opinions when it comes to aesthetics.
2. Zajonc 1968.
3. Frederick and Loewenstein 1999.
4. Rajecki 1974.
5. Monahan, Murphy, and Zajonc 2000
6. Montoya et al. 2017.
7. I wrote about this in great detail in Klein 2005, see also the sources cited there.
8. Sarason and Spielberger 1975; Panksepp 1998; Boyd 2020.
9. Boyd 2020; Arenas and Mazanedo 2020; Klein 2005.
10. Hunter and Schellenberg 2011.
11. Ladd and Gabrieli 2015.
12. Winkielman and Cacioppo 2001.
13. Reber, Winkielman, and Schwarz 1998.
14. Whittlesea 1993.
15. If stimulus processing in the brain is delayed by disruptive tasks, forcing the brain to process familiar and unfamiliar stimuli equally quickly, the mere exposure effect no longer occurs (Leynes and Addante 2016).
16. James 1899.
17. Yin and Knowlton 2006; Wood 2017.
18. Dickinson 1985.
19. Neal et al. 2011.
20. Armitage 2005.
21. Stables et al. 2002.
22. Quinn et al. 2010; Wood and Rünger 2016.
23. Disenis and Bellenoux 2010; Borys et al. 2013

24 statista.com/statistics/270229/usage-duration-of-social-networks-by-country.
25 Connell, Adam. 'Social Media Statistics for 2025: Usage Trends & Data to Know.' AdamConnell.me, 2025. adamconnell.me/social-media-statistics.
26 Grüning, Riedel and Lorenz-Spreen 2023. More information about this app can be found at one-sec.app.
27 Armellino et al. 2012.

Third Illusion: It's always turned out fine before

1 Plato 1980. The original Ancient Greek text uses the words ἀρετή and ἀρεταῖος. These terms have traditionally been translated with the outdated terms 'virtue' and 'virtuous'. What is meant is behaviour that is appropriate and ethical in a given situation.
2 de.statista.com/statistik/daten/studie/1097636/umfrage/zukunftserwartung-fuer-kuenftige-generationen-in-deutschland.
3 This figure is higher than in almost every other European country. The survey was carried out by the Bertelsmann Foundation in November 2019. eupinions.eu/de/text/the-optimism-gap.
4 Kuper-Smith et al. 2021; Debbeler, Schupp, and Renner 2021; Sharot 2012.
5 Rogers et al. 2017.
6 Gallagher, Lopez, and Pressman 2013; Sharot 2011.
7 Sharot, Korn, and Dolan 2011.
8 Pronin, Gilovich, and Ross 2004.
9 McKay and Dennett 2009.
10 Johnson and Fowler 2011.
11 Gifford et al. 2009; Schultz et al. 2014.
12 guvh.de/presse-medien/artikel/2020/02/kugelschreiber.php.
13 Hertwig and Wulff 2022.

14 Hallmann et al. 2017.
15 Van der Heiden, 2023; Watts et al. 2021; zdf.de/nachrichten/politik/klima-eu-umweltagentur-luftqualitaet-feinstaub-100.html

Fourth Illusion: Knowledge is power
1 Hebra 1847.
2 Nevins 2011.
3 Sharot 2017.
4 For a detailed review of the literature, see Sharot 2017.
5 Ecker et al. 2022.
6 Kappes et al. 2020.
7 Gordon 2008.
8 Greenwald 1980.
9 Campbell and Kay 2014.
10 Nuland 2003.
11 Godlee 1924.

Fifth Illusion: Freedom is the answer to everything
1 I wrote extensively on this in my previous works.
2 Bursztyn et al. 2023 argue that users are trapped not only in their behaviour but in social networks in general.
3 European Investment Bank 2020.
4 transportenvironment.org/discover/ryanair-europes-7th-biggest-carbon-polluter-last-year-aviation-emissions-continued-grow.
5 *Datenreport 2013*; Wurthmann 2022, gesis.org/fileadmin/upload/dienstleistung/daten/umfragedaten/allbus/Frageprogramm/Postmaterialismus.html.
6 Inglehart 2015.
7 Klein 2005; 2018.
8 Delhey 2010.
9 kba.de/DE/Statistik/Fahrzeuge/Bestand/bestand_node.html.

10 The only significant decline was during the Covid-19 pandemic. Grömling 2021.
11 Becker, Gerding, and Popp 2014.
12 Klein 2005; 2018.
13 Bursztyn et al. 2023.
14 Sauer 2024.
15 European Investment Bank. 2020; eib.org/en/surveys/2nd-climate-survey/index.htm.
16 Nemet and Johnson 2010; Graham et al. 2019; Rapeli and Koskimaa 2022; Bardsley et al. 2022.
17 Chaikumbung 2023.
18 Hornsey et al. 2016.

Sixth Illusion: We only want the best
1 Machiavelli 1908.
2 Pendergrast 2000.
3 Oliver 1987.
4 Schulte 2021.
5 Sharot 2017.
6 Behavioural economists call this 'status quo bias', which is equivalent to the endowment effect.
7 Edgerton 1992.
8 Bailey and Aunger 1989.
9 McKie 2016.
10 IRENA 2023.
11 Palzer and Henning 2014.
12 This is calculated on the basis of the difference between the necessary investment in renewable energy sources and the resulting economic savings compared to a fossils-based energy supply, as set out in Stern's report.
13 Federal Audit Office (Bundesrechnungshof) 2024.
14 Dubow and Childs 1998.
15 Lang, Weir, and Pearson-Merkowitz 2021.
16 Börjesson, Eliasson, and Hamilton 2016.

Seventh Illusion: Ideologies are obsolete

1. This chapter is partly based on an essay I published in *Zeit Magazin*, issue 22, 2018, under the title 'How Ideology Is Created'.
2. Asch 1951.
3. Zmigrod 2022.
4. Pfündel, Stichs, and Tanis 2021.
5. Weber 2019; Wagner et al. 2006; *Annual Report on Migration and Integration* 2004; 2003; Dylong and Uebelmesser 2024
6. zeit.de/wirtschaft/2023-11/migration-fachkraeftemangel-pull-faktoren-asylpolitik/komplettansicht; tagesschau.de/wirtschaft/wirtschaftsweise-schnitzer-zuwanderung-fachkraefte-100.html.
7. Henrich 2017.
8. Dorling 2016.
9. Carnes and Lupu 2017; Maniam 2016
10. NBC News 2024; Wolf et al. 2024.
11. Adorno 1950.
12. Hatemi et al. 2014
13. Zmigrod, Rentfrow, and Robbins 2018; Zmigrod and Tsakiris 2021; Zmigrod et al. 2021.
14. The answer is 'cheese'.
15. NBC News 2024.
16. Zmigrod and Tsakiris 2021; Rollwage, Dolan, and Fleming 2018.
17. Zmigrod, Rentfrow, and Robbins 2018; Zmigrod et al. 2021; Grover 2021.
18. Uddin 2021; Braem and Egner 2018.
19. Németh et al. 2024.
20. Brandt et al. 2021.
21. The Covid-19 pandemic had a similar effect. During the first lockdown, British social psychologists surveyed

3,000 people in the United Kingdom and Ireland. They found that the greater a respondent's fear of the virus was, the more likely they were to agree with statements such as, 'God's laws about abortion, pornography, and marriage must be strictly followed before it is too late', 'Immigrants harm (our) culture', and 'What our country needs most is discipline'. It might be argued that older people, for example, are naturally more concerned for their health than younger people, and that they are generally more conservative, authoritarian, and nationalistic in their views. But when the researchers adjusted their figures for factors such as age, education level, and sex, they still found a strong correlation between the fear of Covid and an authoritarian worldview (Hartman et al. 2021).
22 Van Prooijen, Douglas, and De Inocencio 2018.
23 There is also a wealth of literature on the effect of religious practice on elementary cognitive processes. See, for example, Good, Inzlicht, and Larson 2015.
24 Arendt 2017.
25 Caplan 2011.
26 Fitzgerald 1945.

Escaping addiction
1 Wolf-Pommrich 1956.
2 Institut für Demoskopie Allensbach / Allensbach Foundation for Public Opinion Research 2008.
3 Wald, Nicholas, and Richard Doll. 'Smoking and Other Risk Factors for Stroke.' *British Medical Journal* 321, no. 7257 (August 5, 2000): 325–29.
4 These figures include the consumption of e-cigarettes, see, e.g., dhs.de/suechte/tabak/zahlen-daten-fakten.
5 U.S. Department of Health and Human Services 2000.
6 Porter 1971.

7   Brandt 2009.
8   Andreas 2019.
9   Bernays 1928.
10 O'Keefe and Pollay 1996.
11 Lickint 1930; Hoffman 1931.
12 U.S. Department of Health and Human Services 2000.
13 Marshall 2016.
14 Ibid.
15 Ibid.
16 Appiah 2011.
17 Ling and Glantz 2004.
18 Marshall 2016.
19 Monroe 1998.
20 CBS-TV 1978.
21 Pacheco 2012; 2013b.
22 Glantz and Balbach 2000.
23 Albers et al. 2007; Hamilton, Biener, and Brennan 2007; Pacheco 2012; 2013; Larsen 2019.
24 Cutler and Glaeser 2006.
25 debra-study.info/wp-content/uploads/2022/12/Factsheet-09-v3.pdf, 'Abbildung 3'.

The great liberation

1   Blanc 2022.
2   Ibid.
3   *'Feuille du salut public'* 1793.
4   In the preface to her *Declaration*.
5   Christakis and Fowler 2009.
6   Grossmann et al. 2023.
7   Godwin 1798.
8   Lund 2012.
9   Gordon 2014.
10 Drescher 2009.
11 Davis 2006.

12 Osterhammel 2014.
13 Drescher 2002.
14 museumcollections.hullcc.gov.uk/collections/search-results/display.php?irn=48586.
15 Davis 2006.

An instruction manual for a better world
1 Judt 2006, p. 268.
2 Klein 2021.
3 Eickelmann et al. 2024.
4 Olbrisch 2024.
5 Henley 2020.
6 E.g. Lessenski 2023.
7 Hertwig and Grüne-Yanoff 2017.
8 Nisbett et al. 1987.
9 Ludolph and Schulz 2018; Aczel et al. 2015; Yoon, Scopelliti, and Morewedge 2021.
10 Morewedge et al. 2015.
11 Kahneman, Sibony, and Sunstein 2021.
12 Fong, Krantz, and Nisbett 1986; Nisbett et al. 1987; Gigerenzer and Selten 2002; Gigerenzer 2014.
13 Morewedge et al. 2015; Sellier, Scopelliti, and Morewedge 2019.
14 Vlasceanu et al. 2024; Nisa et al. 2019.
15 Abrahamse 2019; Allcott and Rogers 2014.
16 Pajula et al. 2021.
17 Otto et al. 2020.
18 Poore and Nemecek 2018.
19 Grossmann et al. 2023.
20 Monbiot 2022.
21 Kanter 1977; Dahlerup 1988; Marwell and Oliver 1993; Centola et al. 2018; Otto et al. 2020
22 Bollinger and Gillingham 2012; Noonan, Hsieh, and Matisoff 2013; 2015; Palm 2017; Abrahamse 2019.

23 Murase and Hilbe 2024.
24 Fowler 2005; Christakis and Fowler 2010.
25 Saint-Exupery 2000.
26 Barnes 1993.
27 Rothman and Salovey 1997; Rothman et al. 2006; Latimer et al. 2008; Gallagher and Updegraff 2012.
28 Lunn et al. 2020.
29 Sapolsky 1994.
30 Happiness research, the quantitative study of subjective life satisfaction, which has now become routine and widespread, repeatedly delivers the same result: once a certain level of income has been reached, a further increase in wealth has only a minimally positive effect on subjective wellbeing — if any at all. Other factors then become important in defining what makes a good life. They include time spent with family and friends, good health, low stress, a healthy natural environment, and the opportunity to develop one's talents and experience interesting new things. See Klein 2005, 2019 and the literature cited there.
31 An example of this is the attempt to introduce a four-day working week in Iceland (Villegas and Knowles 2021). There was a significant increase in productivity and satisfaction among workers, who were also found to be in better physical and mental health. The trials were so successful that more than 85 per cent of Icelandic employees now work shorter hours. Working a four-day week also reduces employees' carbon footprint by at least 20 per cent (Chiu 2022; Knight, Rosa, and Schor 2013).
32 Social scientists call this phenomenon a 'double dividend'. It has been confirmed consistently by many large-scale studies, not only in Europe, and America, but also in China, India, South Africa, and Vietnam.

Jackson 2005; Zawadzki, Steg, and Bouman 2020; Schmitt et al. 2018; Capstick et al. 2022; Nguyen et al. 2024; Blackburn et al. 2024.
33 Havel 1991.

Bibliography

Abrahamse, Wokje. 2019. *Encouraging Pro-environmental Behaviour: What Works, What Doesn't, and Why*. Cambridge, Massachusetts: Academic Press.

Acemoglu, Daron. 2021. 'AI's Future Doesn't Have to Be Dystopian'. *Boston Review*, 20 May 2021; bostonreview.net/forum/ais-future-doesnt-have-to-be-dystopian.

Acquisti, Alessandro and Ralph Gross. 2006. 'Imagined Communities: Awareness, Information Sharing, and Privacy on the Facebook'. In *International Workshop on Privacy Enhancing Technologies*, 36–58. Springer.

Aczel, Balazs, Bence Bago, Aba Szollosi, Andrei Foldes, and Bence Lukacs. 2015. 'Is It Time for Studying Real-life Debiasing? Evaluation of the Effectiveness of an Analogical Intervention Technique'. *Frontiers in Psychology* 6 (August); doi.org/10.3389/fpsyg.2015.01120.

ADAC [General German Automobile Club]. 2023. 'Kostenvergleich: E-Auto, Benziner oder Diesel?' ['Cost Comparison: EV, petrol or diesel?'] 28 April 2023; dac.de/rund-ums-fahrzeug/auto-kaufen-verkaufen/autokosten/elektroauto-kostenvergleich.

Adorno, Theodor W., Else Frenkel-Brunswik, Daniel J. Levinson, and R. Nevitt Sanford 1950. *The Authoritarian*

Personality: Studies in Prejudice. New York: Harper & Brother.

Aiken, Carolyn and Scott Keller. 2009. 'The Irrational Side of Change Management'. *McKinsey Quarterly* 2 (10): 100–109.

Albers, Alison B, Michael Siegel, Debbie M Cheng, Lois Biener, and Nancy A Rigotti. 2007. 'Effect of Smoking Regulations in Local Restaurants on Smokers' Anti-Smoking Attitudes and Quitting Behaviours'. *Tobacco Control* 16 (2): 101–6; doi.org/10.1136/tc.2006.017426.

Allcott, Hunt and Todd Rogers. 2014. 'The Short-Run and Long-Run Effects of Behavioral Interventions: Experimental Evidence from Energy Conservation'. *American Economic Review* 104 (10): 3003–37; doi.org/10.1257/aer.104.10.3003.

Andreas, Peter. 2019. 'Drugs and War: What Is the Relationship?' *Annual Review of Political Science* 22: 57–73; doi.org/10.1146/annurev-polisci-051017-103748.

Appiah, Kwame A. 2011. *The Honor Code: How Moral Revolutions Happen*. Reprint Edition. New York: WW Norton.

Arenas, M. C. and C. Mazanedo. 2020. 'Novelty Seeking'. In *Encyclopedia of Personality and Individual Differences*, Virgil Zeigler-Hill and Todd K. Shackelford (eds). Cham: Springer International Publishing; doi.org/10.1007/978-3-319-24612-3_672.

Arendt, Hannah. 2017. *The Origins of Totalitarianism*. 1st ed. London: Penguin Classics.

Armellino, Donna, Erfan Hussain, Mary Ellen Schilling, William Senicola, Ann Eichorn, Yosef Dlugacz, and Bruce F. Farber. 2012. 'Using High-Technology to Enforce Low-Technology Safety Measures: The Use of Third-party Remote Video Auditing and Real-time Feedback in Healthcare'. *Clinical Infectious Diseases* 54 (1): 1–7; doi.org/10.1093/cid/cir773.

Armitage, Christopher J. 2005. 'Can the Theory of Planned Behavior Predict the Maintenance of Physical Activity?' *Health Psychology* 24 (3): 235; psycnet.apa.org/fulltext/2005-04818-001.html.

Asch, S. 1951. 'Effects of Group Pressure the Modification and Distortion of Judgment'. In *Groups, Leadership and Men*. Pittsburgh: Carnegie Press.

Bailey, Robert C. and Robert Aunger. 1989. 'Net Hunters vs. Archers: Variation in Women's Subsistence Strategies in the Ituri Forest'. *Human Ecology* 17 (3): 273–97; jstor.org/stable/4602926.

Bardsley, Nicholas, Graziano Ceddia, Rachel McCloy, and Simone Pfuderer. 2022. 'Why Economic Valuation Does Not Value the Environment: Climate Policy as Collective Endeavour'. *Environmental Values* 31 (3): 277–93; doi.org/10.3197/096327121X16081160834740.

Barnes, Julian. 1993. *Flaubert's Parrot*. New York: Random House.

Becker, Sven, Jonas Gerding and Maximilian Popp. 2014. 'Generation Ich' ['The Me Generation']. *Der Spiegel*, 26 October 2014, Politics Section (paywall); spiegel.de/politik/generation-ich-a-d5f69ca3-0002-0001-0000-000129976908.

Bernays, Edward L. 1928. *Propaganda*. New York: Horace Liveright.

——. 1969. *The Engineering of Consent*. Oklahoma: University of Oklahoma Press.

Blackburn, Rebecca, Zoe Leviston, Iain Walker, and Ashley Schram. 2024. 'Could a Minimalist Lifestyle Reduce

Carbon Emissions and Improve Wellbeing? A Review of Minimalism and Other Low Consumption Lifestyles'. *WIREs Climate Change* 15 (2): e865; doi.org/10.1002/wcc.865.

Blanc, Olivier. 2022. *Olympe de Gouges: De la déclaration des droits de la femme et de la citoyenne à la guillotine*. [*On the Declaration of the Rights of Women and of Citizens at the Guillotine*]. Paris: Tallandier.

Bollinger, Bryan and Kenneth Gillingham. 2012. 'Peer Effects in the Diffusion of Solar Photovoltaic Panels'. *Marketing Science* 31 (6): 900–912; doi.org/10.1287/mksc.1120.0727.

Börjesson, Maria, Jonas Eliasson, and Carl Hamilton. 2016. 'Why Experience Changes Attitudes to Congestion Pricing: The Case of Gothenburg'. *Transportation Research Part A: Policy and Practice* 85 (March): 1–16; doi.org/10.1016/j.tra.2015.12.002.

Borys, J.-M., L. Valdeyron, E. Levy, J. Vinck, D. Edell, L. Walter, H. Ruault du Plessis, P. Harper, P. Richard, and A. Barriguette. 2013. 'EPODE—A Model for Reducing the Incidence of Obesity and Weight-Related Comorbidities'. *European Endocrinology* 9 (2): 116–20; pmc.ncbi.nlm.nih.gov/articles/PMC6003578.

Boyd, Patrick. 2020. 'Openness'. In *Encyclopedia of Personality and Individual Differences*, Virgil Zeigler-Hill and Todd K. Shackelford (eds). Cham: Springer International Publishing; doi.org/10.1007/978-3-319-24612-3_672.

Braem, Senne and Tobias Egner. 2018. 'Getting a Grip on Cognitive Flexibility'. *Current Directions in Psychological Science* 27 (6): 470–76; doi.org/10.1177/0963721418787475.

Brandt, Allan M. 2009. *The Cigarette Century: The Rise, Fall, and Deadly Persistence of the Product That Defined America*. New York: Basic Books.

Brandt, Mark J., Felicity M. Turner-Zwinkels, Beste Karapirinler, Florian Van Leeuwen, Michael Bender, Yvette van Osch, and Byron Adams. 2021. 'The Association Between Threat and Politics Depends on the Type of Threat, the Political Domain, and the Country'. *Personality and Social Psychology Bulletin* 47 (2): 324–43; doi.org/10.1177/0146167220946187.

Bursztyn, Leonardo, Benjamin R. Handel, Rafael Jimenez, and Christopher Roth. 2023. 'When Product Markets Become Collective Traps: The Case of Social Media'. *National Bureau of Economic Research*; nber.org/papers/w31771.

Campbell, T. and A. Kay. 2014. 'Solution Aversion: On the Relation Between Ideology and Motivated Disbelief'. *Journal of Personality and Social Psychology* 107 (5): 809–24.

Caplan, Bryan. 2011. 'The Ideological Turing Test'. *Econlib*. 20 June 2011; econlib.org/archives/2011/06/the_ideological.html.

Capstick, Stuart, Nicholas Nash, Lorraine Whitmarsh, Wouter Poortinga, Paul Haggar, and Adrian Brügger. 2022. 'The Connection between Subjective Wellbeing and Pro-environmental Behaviour: Individual and Cross-national Characteristics in a Seven-country Study'. *Environmental Science & Policy* 133 : 63–73; sciencedirect.com/science/article/pii/S1462901122000776.

Carleton, Tamma, Amir Jina, Michael Delgado, Michael Greenstone, Trevor Houser, Solomon Hsiang, Andrew Hultgren et al. 2022. 'Valuing the Global Mortality

Consequences of Climate Change Accounting for Adaptation Costs and Benefits'. *The Quarterly Journal of Economics* 137 (4): 2037–2105; doi.org/10.1093/qje/qjac020.

Carnes, Nicholas and Noam Lupu. 2017. 'Analysis: It's Time to Bust the Myth: Most Trump Voters Were Not Working Class'. *Washington Post*, 5 June 2017; washingtonpost.com/news/monkey-cage/wp/2017/06/05/its-time-to-bust-the-myth-most-trump-voters-were-not-working-class.

CBS-TV. 1978. Magazine with Betty Carnes, Scottsdale, Arizona; archive.org/details/tobacco_qhw27a00.

Centola, Damon, Joshua Becker, Devon Brackbill, and Andrea Baronchelli. 2018. 'Experimental Evidence for Tipping Points in Social Convention'. *Science* 360 (6393): 1116–19; doi.org/10.1126/science.aas8827.

Chaikumbung, Mayula. 2023. 'The Effects of Institutions and Cultures on People's Willingness to Pay for Climate Change Policies: A Meta-regression Analysis'. *Energy Policy* 177 (June):113513; doi.org/10.1016/j.enpol.2023.113513.

Chiu, Allyson. 2022. 'How a Four-Day Workweek Could Be Better for the Climate'. *Washington Post*, 8 August 2022; ashingtonpost.com/climate-solutions/2022/08/08/4-day-workweek-environment.

Christakis, Nicholas A. and James H. Fowler. 2009. *Connected: The Surprising Power of Our Social Networks and How They Shape Our Lives*. New York: Little, Brown and Co; archive.org/details/connectedsurpris00chri.

———. 2010. *Connected: The Surprising Power of Our Social Networks and How They Shape Our Lives*. New York: Little, Brown and Co.

Clark, Andy. 2019. *Surfing Uncertainty: Prediction, Action, and the Embodied Mind*. Reprint Edition. Oxford New York: Oxford University Press.

Cook, Benjamin, K. J. Anchukaitis, Jed O. Kaplan, M. J. Puma, M. Kelley, and D. Gueyffier. 2012. 'Pre-Columbian Deforestation as an Amplifier of Drought in Mesoamerica'. *Geophysical Research Letters* 39 (16).

Cook, Benjamin, Jason E. Smerdon, Edward R. Cook, A. Park Williams, Kevin J. Anchukaitis, Justin S. Mankin, Kathryn Allen, Laia Andreu-Hayles, Toby R. Ault, and Soumaya Belmecheri. 2022. 'Megadroughts in the Common Era and the Anthropocene'. *Nature Reviews Earth & Environment* 3 (11): 741–57.

Cutler, David M. and Edward L. Glaeser. 2006. 'Why Do Europeans Smoke More than Americans?' National Bureau of Economic Research Cambridge, Mass., USA; nber.org/papers/w12124.

Dahlerup, Drude. 1988. 'From a Small to a Large Minority: Women in Scandinavian Politics'. *Scandinavian Political Studies* 11 (4): 275–98; doi.org/10.1111/j.1467-9477.1988.tb00372.x.

Davis, David Brion. 2006. *Inhuman Bondage: The Rise and Fall of Slavery in the New World*. New York: Oxford University Press.

Debbeler, Luka J., Harald T. Schupp, and Britta Renner. 2021. 'Pessimistic Health and Optimistic Wealth Distributions Perceptions in Germany and the UK: Evidence from an Online-Survey'. *BMC Public Health* 21 (1): 1306; doi.org/10.1186/s12889-021-11355-x.

Dechezleprêtre, Antoine, Adrien Fabre, Tobias Kruse, Bluebery Planterose, Ana Sanchez Chico, and Stefanie Stantcheva. 2022. 'Fighting Climate Change:

International Attitudes Toward Climate Policies'. National Bureau of Economic Research.

Delhey, Jan. 2010. 'From Materialist to Post-Materialist Happiness? National Affluence and Determinants of Life Satisfaction in Cross-National Perspective'. *Social Indicators Research* 97 (1): 65–84; doi.org/10.1007/s11205-009-9558-y.

Dickinson, Anthony. 1985. 'Actions and Habits: The Development of Behavioural Autonomy'. *Philosophical Transactions of the Royal Society of London B, Biological Sciences* 308 (1135): 67–78; doi.org/10.1098/rstb.1985.0010.

Disenis, Caroline and Ellenore Bellenoux. 2010. 'Le dossier de présentation des premiers résultats d'Epode' ['Presentation report of the initial results of EPODE']. Paris; banquedesterritoires.fr/sites/default/files/ra/Le%20dossier%20de%20pr%C3%A9sentation%20des%20premiers%20r%C3%A9sultats%20d%27Epode.pdf.

Dorling, Danny. 2016. 'Brexit: The Decision of a Divided Country'. *The British Medical Journal*, July, i3697; doi.org/10.1136/bmj.i3697.

Drescher, Seymour. 2002. *The Mighty Experiment: Free Labor versus Slavery in British Emancipation.* New York: Oxford University Press.

———. 2009. *Abolition: A History of Slavery and Antislavery.* Cambridge: Cambridge University Press.

Dubow, Joel S. and Nancy M. Childs. 1998. 'New Coke, Mixture Perception, and the Flavor Balance Hypothesis'. *Journal of Business Research* 43 (3): 147–55; sciencedirect.com/science/article/abs/pii/S0148296397002208.

Dylong, Patrick and Silke Uebelmesser. 2024. 'Biased Beliefs about Immigration and Economic Concerns:

Evidence from Representative Experiments'. *Journal of Economic Behavior & Organization* 217: 453–82; sciencedirect.com/science/article/pii/S0167268123004250.

Ecker, Ullrich K. H., Stephan Lewandowsky, John Cook, Philipp Schmid, Lisa K. Fazio, Nadia Brashier, Panayiota Kendeou, Emily K. Vraga, and Michelle A. Amazeen. 2022. 'The Psychological Drivers of Misinformation Belief and Its Resistance to Correction'. *Nature Reviews Psychology* 1 (1): 13–29; doi.org/10.1038/s44159-021-00006-y.

Edgerton, Robert. 1992. *Sick Societies: Challenging the Myth of Primitive Harmony*. New York: Free Press.

Eickelmann, Birgit, Nadine Fröhlich, Gianna Casamassima, and Kerstin Drossel. 2024. *ICILS 2023 im Überblick. Zentrale Ergebnisse, Entwicklungen über ein Jahrzehnt und mögliche Entwicklungsperspektiven.* [*A Summary of ICILS in 2023. Central Results, Developments Over a Decade, and Possible Perspectives for Future Development.*] Waxmann Verlag GmbH; doi.org/10.31244/9783830999416.

Engels, Barbara and Mara Grunewald. 2017. 'Das Privacy Paradox: Digitalisierung versus Privatsphäre.' ['The Privacy Paradox: Computerisation versus Privacy'.] 27. 2017. Kurzberichte Des Instituts Der Deutschen Wirtschaft. Cologne. [Cologne Institute for Economic Research Short Reports].

European Commission. 2024. *Eurobarometer 538 Climate Change*. Brussels: European Commission.

European Investment Bank. 2020. *The EIB Climate Survey 2019–2020: How Citizens Are Confronting the Climate Crisis and What Actions They Expect from Policymakers and Businesses*. LU: Publications Office; data.europa.eu/doi/10.2867/653713.

Evans, Nicholas P., Thomas K. Bauska, Fernando Gázquez-Sánchez, Mark Brenner, Jason H. Curtis, and David A. Hodell. 2018. 'Quantification of Drought during the Collapse of the Classic Maya Civilization'. *Science* 361 (6401): 498–501; doi.org/10.1126/science.aas9871.

Expert Council on Integration and Migration *Annual Report on Migration and Integration*. 2004. Nuremberg.

Farmer, Steve, Richard Sproat, and Michael Witzel. 2004. 'The Collapse of the Indus-Script Thesis: The Myth of a Literate Harappan Civilization'. *Electronic Journal of Vedic Studies* 11 (2): 19–57; doi.org/10.11588/ejvs.2004.2.620.

Federal Audit Office (Bundesrechnungshof). 2024. *Bericht nach § 99 BHO zur Umsetzung der Energiewende im Hinblick auf die Versorgungssicherheit, Bezahlbarkeit und Umweltverträglichkeit der Stromversorgung*. Bonn: Bundesrechnungshof. [*Report according to § 99 of the Federal Budget Code, on the Implementation of the Energy Transition with Regard to the Supply Security, Affordability and Environmental Sustainability of the Electricity Supply*, summarised in English at: bundesrechnungshof.de/SharedDocs/Kurzmeldungen/EN/2024/energiewende-en.html.]

Festinger, Leon, Henry W. Riecken, and Stanley Schachter. 1956. *When Prophecy Fails*. Minneapolis: University of Minnesota Press; ia802802.us.archive.org/4/items/pdfy-eDNpDzTy_dR1b0iB/Festinger-Riecken-Schachter-When-Prophecy-Fails-1956.pdf.

'*Feuille du salut public*'. 1793. *RetroNews*; retronews.fr/journal/feuille-du-salut-public/17-novembre-1793/1639/2855785/1.

Fitzgerald, Francis Scott. 1945. *The Crack-Up*. New York: New Directions.

FitzGerald, Thomas H. B., Raymond J. Dolan, and Karl J. Friston. 2014. 'Model Averaging, Optimal Inference, and Habit Formation'. *Frontiers in Human Neuroscience* 8; frontiersin.org/articles/10.3389/fnhum.2014.00457.

Fong, Geoffrey T., David H. Krantz, and Richard E. Nisbett. 1986. 'The Effects of Statistical Training on Thinking about Everyday Problems'. *Cognitive Psychology* 18 (3): 253–92; sciencedirect.com/science/article/pii/0010028586900010.

Fowler, J. H. 2005. 'Turnout in a Small World'. In Zuckerman A.S., (ed.) *The Social Logic of Politics*. Philadelphia: Temple University Press.

Frederick, Shane and George Loewenstein. 1999. 'Chapter 16: Hedonic Adaptation'. *Well-Being. The Foundations of Hedonic Psychology*, edited by D. Kahneman and E. Diener, 302–29. New York: Russell Sage Foundation

Friston, Karl. 2018. 'Does Predictive Coding Have a Future?' *Nature Neuroscience* 21 (8): 1019–21; doi.org/10.1038/s41593-018-0200-7.

Gallagher, Kristel M. and John A. Updegraff. 2012. 'Health Message Framing Effects on Attitudes, Intentions, and Behavior: A Meta-Analytic Review'. *Annals of Behavioral Medicine* 43 (1): 101–16; academic.oup.com/abm/article-abstract/43/1/101/4563944.

Gallagher, Matthew W., Shane J. Lopez, and Sarah D. Pressman. 2013. 'Optimism is Universal: Exploring the Presence and Benefits of Optimism in a Representative Sample of the World'. *Journal of Personality* 81 (5): 429–40; doi.org/10.1111/jopy.12026.

Gifford, Robert, Leila Scannell, Christine Kormos, Lidia Smolova, Anders Biel, Stefan Boncu, Victor Corral et al. 2009. 'Temporal Pessimism and Spatial Optimism in Environmental Assessments: An 18-Nation Study'. *Journal of Environmental Psychology* 29 (1): 1–12; doi.org/10.1016/j.jenvp.2008.06.001.

Gigerenzer, Gerd. 2014. *Risk Savvy: How to Make Good Decisions*. New York: Viking.

Gigerenzer, Gerd and Reinhard Selten. 2002. *Bounded Rationality: The Adaptive Toolbox*. Cambridge, Massachusetts: MIT Press.

Gill, Richardson Benedict. 2000. *The Great Maya Droughts: Water, Life, and Death*. Albuquerque: University of New Mexico Press.

Glantz, Stanton A. and Edith D. Balbach. 2000. *Tobacco War: Inside the California Battles*. Oakland: University of California Press.

Godlee, Rickman John. 1924. *Lord Lister*. 3rd edition, revised. Oxford: Clarendon Press.

Godwin, William. 1798. *Memoirs of the Author of a Vindication of the Rights of Woman*. 2nd edition, corrected. Eighteenth Century Collection Online. London: Printed for J. Johnson.

Good, Marie, Michael Inzlicht, and Michael J. Larson. 2015. 'God Will Forgive: Reflecting on God's Love Decreases Neurophysiological Responses to Errors'. *Social Cognitive and Affective Neuroscience* 10 (3): 357–63; doi.org/10.1093/scan/nsu096.

Gordon, Lyndall. 2014. *Vindication: A Life of Mary Wollstonecraft*. London: Hachette.

Gordon, Richard. 2008. *Great Medical Disasters*. New Edition. Cornwall: House of Stratus.

Graham, H., S. de Bell, N. Hanley, S. Jarvis, and P. C. L. White. 2019. 'Willingness to Pay for Policies to Reduce Future Deaths from Climate Change: Evidence from a

British Survey'. *Public Health* 174 (September): 110–17; doi.org/10.1016/j.puhe.2019.06.001.

Greenwald, Anthony G. 1980. 'The Totalitarian Ego: Fabrication and Revision of Personal History'. *American Psychologist* 35 (7): 603; psycnet.apa.org/record/1980-24373-001.

Grömling, Michael. 2021. *Private Consumption in Germany. IW Trends*, No. 2; doi.org/10.2373/1864-810X.21-02-01.

Grossmann, Igor, Amanda Rotella, Cendri A. Hutcherson, Konstantyn Sharpinskyi, Michael E. W. Varnum, Sebastian Achter, Mandeep K. Dhami et al. 2023. 'Insights into the Accuracy of Social Scientists' Forecasts of Societal Change'. *Nature Human Behaviour* 7 (4): 484–501; doi.org/10.1038/s41562-022-01517-1.

Grover, Natalie. 2021. 'People with Extremist Views Less Able to Do Complex Mental Tasks, Research Suggests'. *The Guardian*, 22 February 2021, Science section; theguardian.com/science/2021/feb/22/people-with-extremist-views-less-able-to-do-complex-mental-tasks-research-suggests.

Grüning, David J., Frederik Riedel, and Philipp Lorenz-Spreen. 2023. 'Directing Smartphone Use through the Self-nudge App One Sec'. *Proceedings of the National Academy of Sciences* 120 (8): e2213114120; doi.org/10.1073/pnas.2213114120.

Haidt, Jonathan. 2024. *The Anxious Generation: How the Great Rewiring of Childhood Is Causing an Epidemic of Mental Illness*. New York: Penguin Press.

Hallmann, Caspar A., Martin Sorg, Eelke Jongejans, Henk Siepel, Nick Hofland, Heinz Schwan, Werner Stenmans et al. 2017. 'More than 75 Percent Decline over 27 Years in Total Flying Insect Biomass in Protected Areas'.

PLOS ONE 12 (10): e0185809; doi.org/10.1371/journal.pone.0185809.

Hamilton, W. L., L. Biener, and R. T. Brennan. 2007. 'Do Local Tobacco Regulations Influence Perceived Smoking Norms? Evidence from Adult and Youth Surveys in Massachusetts'. *Health Education Research* 23 (4): 709–22; doi.org/10.1093/her/cym054.

Hartman, Todd K., Thomas V. A. Stocks, Ryan McKay, Jilly Gibson-Miller, Liat Levita, Anton P. Martinez, Liam Mason et al. 2021. 'The Authoritarian Dynamic During the COVID-19 Pandemic: Effects on Nationalism and Anti-Immigrant Sentiment'. *Social Psychological and Personality Science* 12 (7): 1274–85; doi.org/10.1177/1948550620978023.

Hatemi, Peter K., Sarah E. Medland, Robert Klemmensen, Sven Oskarsson, Levente Littvay, Christopher T. Dawes, Brad Verhulst et al. 2014. 'Genetic Influences on Political Ideologies: Twin Analyses of 19 Measures of Political Ideologies from Five Democracies and Genome-Wide Findings from Three Populations'. *Behavior Genetics* 44 (3): 282–94; doi.org/10.1007/s10519-014-9648-8.

Haug, Gerald H., Detlef Gunther, Larry C. Peterson, Daniel M. Sigman, Konrad A. Hughen, and Beat Aeschlimann. 2003. 'Climate and the Collapse of Maya Civilization'. *Science* 299 (5613): 1731–35.

Havel, Václav. 1991. *Disturbing the Peace. A Conversation with Karel Hvíždala.* New York: Knopf Doubleday.

Hebra, Ferdinand von. 1847. 'Höchst wichtige Erfahrungen über die Ätiologie der in Gebäranstalten epidem. Puerperalfieber.' *Zeitschrift der k. k. Gesellschaft der Ärzte zu Wien* ['Highly Important Experiences Concerning the Aetiology of the Puerperal Fever Epidemic in

Obstetrical Institutes.' *Journal of the Imperial and Royal Society of Physicians in Vienna*]. 4: 242.

Heiden, Matthias van der. 2023. 'Neubestimmung der Prädiktionsintervalle zur Schätzung der hitzebedingten Mortalität'. *Epidemiologisches Bulletin* ['Redefining the Prediction Intervals in Assessments of Heat-Related Mortality'. 23 : 14–16; edoc.rki.de/handle/176904/11178.

Henley, Jon. 2020. 'How Finland Starts Its Fight against Fake News in Primary Schools'. *The Guardian*, 29 January 2020, World News Section; theguardian.com/world/2020/jan/28/fact-from-fiction-finlands-new-lessons-in-combating-fake-news.

Henrich, Joseph. 2017. *The Secret of Our Success: How Culture Is Driving Human Evolution, Domesticating Our Species, and Making Us Smarter*. New ed. Princeton: Princeton University.Press.

Hertig, Elke, Iris Hunger, Irena Kaspar-Ott, Andreas Matzarakis, Hildegard Niemann, Lea Schulte-Droesch, and Maike Voss. 2023. 'Klimawandel und Public Health in Deutschland—Eine Einführung in den Sachstandsbericht Klimawandel und Gesundheit 2023' ['Climate Change and Public Health in Germany—An Introduction to the 2023 Status Report on Climate Change and Health']; doi.org/10.25646/11391.

Hertwig, Ralph and Till Grüne-Yanoff. 2017. 'Nudging and Boosting: Steering or Empowering Good Decisions'. *Perspectives on Psychological Science* 12 (6): 973–86; doi.org/10.1177/1745691617702496.

Hodell, David A., Jason H. Curtis, and Mark Brenner. 1995. 'Possible Role of Climate in the Collapse of Classic Maya Civilization'. *Nature* 375 (6530): 391–94.

Hoffman, Frederick L. 1931. 'Cancer and Smoking Habits'. *Annals of Surgery* 93 (1): 50–67. ncbi.nlm.nih.gov/pmc/articles/PMC1398760/.

Hornsey, Matthew J., Emily A. Harris, Paul G. Bain, and Kelly S. Fielding. 2016. 'Meta-analyses of the Determinants and Outcomes of Belief in Climate Change'. *Nature Climate Change* 6 (6): 622–26.

Hunter, Patrick G. and E. Glenn Schellenberg. 2011. 'Interactive Effects of Personality and Frequency of Exposure on Liking for Music'. *Personality and Individual Differences* 50 (2): 175–79; doi.org/10.1016/j.paid.2010.09.021.

Infratest Dimap. 2024. 'ARD-Deutschland TREND Juli 2024'. infratest-dimap.de. 30 May 2024; infratest-dimap.de/umfragen-analysen/bundesweit/ard-deutschlandtrend/2024/juli/.

Inglehart, Ronald. 2015. *The Silent Revolution: Changing Values and Political Styles Among Western Publics.* Princeton: Princeton University Press.

Institut für Demoskopie Allensbach / Allensbach Foundation for Public Opinion Research. 2008. 'Rauchverbote und Raucher' ['Smoking Bans and Smokers']. 2008/1. Allensbacher Berichte. Allensbach; ifd-allensbach.de/fileadmin/kurzberichte_dokumentationen/prd_0801_01.pdf.

IRENA 2023. *Renewable Power Generation Costs in 2022.* International Renewable Energy Agency.

Jackson, Tim. 2005. 'Live Better by Consuming Less?: Is There a "Double Dividend" in Sustainable Consumption?' *Journal of Industrial Ecology* 9 (1–2): 19–36; doi.org/10.1162/1088198054084734.

James, William. 1899. *Talks to Teachers on Psychology and to*

Students on Some of Life's Ideals. Vol. 12. New York.

Johnson, Dominic D. P. and James H. Fowler. 2011. 'The Evolution of Overconfidence'. *Nature* 477 (7364): 317–20; doi.org/10.1038/nature10384.

Judt, Tony. 2006. *Postwar: A History of Europe Since 1945*. New York: Penguin.

Kahneman, Daniel, Olivier Sibony, and Cass R. Sunstein. 2021. *Noise: A Flaw in Human Judgment*. New York: HarperCollins.

Kaltenberg, Mary, Adam B. Jaffe, and Margie E. Lachman. 2023. 'Invention and the Life Course: Age Differences in Patenting'. *Research Policy* 52 (1): 104629.

Kanter, Rosabeth Moss. 1977. 'Some Effects of Proportions on Group Life: Skewed Sex Ratios and Responses to Token Women'. *American Journal of Sociology* 82 (5): 965–90; doi.org/10.1086/226425.

Kappes, Andreas, Ann H. Harvey, Terry Lohrenz, P. Read Montague, and Tali Sharot. 2020. 'Confirmation Bias in the Utilization of Others' Opinion Strength'. *Nature Neuroscience* 23 (1): 130–37; doi.org/10.1038/s41593-019-0549-2.

Kennett, Douglas J., Sebastian FM Breitenbach, Valorie V. Aquino, Yemane Asmerom, Jaime Awe, James UL Baldini, Patrick Bartlein, Brendan J. Culleton, Claire Ebert, and Christopher Jazwa. 2012. 'Development and Disintegration of Maya Political Systems in Response to Climate Change'. *Science* 338 (6108): 788–91.

Klein, Richard. 1993. *Cigarettes Are Sublime*. Durhaam: Duke University Press.

Klein, Stefan. 2005. *The Science of Happiness: How Our Brains Make Us Happy and What We Can Do to Get Happier*. Melbourne: Scribe Publications.

———. 2018. *Die Ökonomie des Glücks: Warum unsere Gesellschaft neue Ziele braucht* [*The Economics of Happiness: Why Our Society Needs New Goals*]. 1st edition. Berlin: Nicolai Publishing & Intelligence GmbH.

———. 2021. *Wie wir die Welt verändern. Eine kurze Geschichte des menschlichen Geistes* [*How We Change the World: A Short History of the Human Mind*]. Frankfurt am Main: S. Fischer Verlag.

Knight, Kyle W., Eugene A. Rosa, and Juliet B. Schor. 2013. 'Could Working Less Reduce Pressures on the Environment? A Cross-national Panel Analysis of OECD Countries, 1970–2007'. *Global Environmental Change* 23 (4): 691–700; doi.org/10.1016/j.gloenvcha.2013.02.017.

Kuper-Smith, Benjamin J., Lisa M. Doppelhofer, Yulia Oganian, Gabriela Rosenblau, and Christoph W. Korn. 2021. 'Risk Perception and Optimism during the Early Stages of the Covid-19 Pandemic'. *Royal Society Open Science* 8 (11): 210904; doi.org/10.1098/rsos.210904.

Ladd, Sandra L. and John D. E. Gabrieli. 2015. 'Trait and State Anxiety Reduce the Mere Exposure Effect'. *Frontiers in Psychology* 6 (May); doi.org/10.3389/fpsyg.2015.00701.

Lang, Corey, Michael Weir, and Shanna Pearson-Merkowitz. 2021. 'Status Quo Bias and Public Policy: Evidence in the Context of Carbon Mitigation'. *Environmental Research Letters* 16 (5): 054076; doi.org/10.1088/1748-9326/abeeb0.

Larsen, Erik Gahner. 2019. 'Policy Feedback Effects on Mass Publics: A Quantitative Review'. *Policy Studies Journal* 47 (2): 372–94; doi.org/10.1111/psj.12280.

Latimer, Amy E., Tara A. Rench, Susan E. Rivers, Nicole

A. Katulak, Stephanie A. Materese, Lisa Cadmus, Althea Hicks, Julie Keany Hodorowski, and Peter Salovey. 2008. 'Promoting Participation in Physical Activity Using Framed Messages: An Application of Prospect Theory'. *British Journal of Health Psychology* 13 (4): 659–81; doi.org/10.1348/135910707X246186.

Lessenski, Marin. 2023. *Media Literacy Index 2023*. Sofia: Open Society Institute.

Lewine, Harris. 1970. *Good-Bye To All That*. New York: McGraw-Hill.

Leynes, P. Andrew and Richard J. Addante. 2016. 'Neurophysiological Evidence That Perceptions of Fluency Produce Mere Exposure Effects'. *Cognitive, Affective, & Behavioral Neuroscience* 16 (4): 754–67; doi.org/10.3758/s13415-016-0428-1.

Lickint, Fritz. 1930. 'Tabak und Tabakrauch als ätiologischer Faktor des Carcinoms.' *Zeitschrift für Krebsforschung* ['Tobacco and Tobacco Smoke as an Aetiological Factor of Carcinoma'. *Journal of Cancer Research*]. 30 (1): 349–65; doi.org/10.1007/BF01636077.

Ling, Pamela M and Stanton A Glantz. 2004. 'Tobacco Industry Research on Smoking Cessation'. *Journal of General Internal Medicine* 19 (5 Pt 1): 419–26; doi.org/10.1111/j.1525-1497.2004.30358.x.

Lorenz-Spreen, Philipp, Lisa Oswald, Stephan Lewandowsky, and Ralph Hertwig. 2023. 'A Systematic Review of Worldwide Causal and Correlational Evidence on Digital Media and Democracy'. *Nature Human Behaviour* 7 (1): 74–101; doi.org/10.1038/s41562-022-01460-1.

Ludolph, Ramona and Peter J. Schulz. 2018. 'Debiasing Health-Related Judgments and Decision Making: A

Systematic Review'. *Medical Decision Making* 38 (1): 3–13; doi.org/10.1177/0272989X17716672.

Lund, Hannah Lotte. 2012. 'Der Berliner ›jüdische Salon‹ um 1800: Emanzipation in der Debatte' in *Der Berliner »jüdische Salon« um 1800*. ['The "Jewish Salon" in Berlin Around the Year 1800: The Debate Over Emancipation' in *The "Jewish Salon" in Berlin Around the Year 1800*]. Berlin: De Gruyter; doi.org/10.1515/9783110271744.

Lunn, Pete, Cameron Belton, Ciarán Lavin, Féidhlim McGowan, Shane Timmons, and Deidre Robertson. 2020. 'Using Behavioural Science to Help Fight the Coronavirus'. 656. ESRI Working Paper. Dublin: The Economic and Social Research Institute.

Machiavelli, Niccolò. 1908. *The Prince*. Trans. W. K. Marriott. London: J.M. Dent & Sons.

Maier, Franz Georg. 1968. *Die Verwandlung der Mittelmeerwelt*. Fischer-Weltgeschichte 9. [*The Transformation of the Mediterranean World*. Fischer World History Vol. 9].

Marshall, Thomas R. 2016. *Public Opinion, Public Policy, and Smoking: The Transformation of American Attitudes and Cigarette Use, 1890–2016*. Lanham: Lexington Books.

Marwell, Gerald and Pamela Oliver. 1993. *The Critical Mass in Collective Action: a Micro-Social Theory*. Cambridge (England) and New York: Cambridge University Press.

McIntosh, Jane R. 2007. *The Ancient Indus Valley: New Perspectives*. New York: Bloomsbury Publishing.

McKay, Ryan T. and Daniel C. Dennett. 2009. 'The evolution of misbelief'. *Behavioral and Brain Sciences* 32 (6): 493–510; cambridge.org/core/journals/behavioral-and-brain-sciences/article/evolution-of-misbelief/F817ED84D1FFD8650D61FB61301203E8.

McKie, Robin. 2016. 'Nicholas Stern: Cost of Global Warming "Is Worse than I Feared".' *The Observer*, 6 November 2016, Environment Section; theguardian.com/environment/2016/nov/06/nicholas-stern-climate-change-review-10-years-on-interview-decisive-years-humanity.

Medina-Elizalde, Martín and Eelco J. Rohling. 2012. 'Collapse of Classic Maya Civilization Related to Modest Reduction in Precipitation'. *Science* 335 (6071): 956–59.

Mithen, Steven, and Emily Black. 2011. *Water, Life and Civilisation: Climate, Environment and Society in the Jordan Valley*. Cambridge: Cambridge University Press.

Monahan, Jennifer L., Sheila T. Murphy, and R. B. Zajonc. 2000. 'Subliminal Mere Exposure: Specific, General, and Diffuse Effects'. *Psychological Science* 11 (6): 462–66; doi.org/10.1111/1467-9280.00289.

Monbiot, George. 2022. *Regenesis: Feeding the World Without Devouring the Planet*. London: Penguin Books.

Monroe, Alan D. 1998. 'Public Opinion and Public Policy, 1980–1993'. *The Public Opinion Quarterly* 62 (1): 6–28; jstor.org/stable/2749715.

Montoya, R. Matthew, Robert S. Horton, Jack L. Vevea, Martyna Citkowicz, and Elissa A. Lauber. 2017. 'A Re-Examination of the Mere Exposure Effect: The Influence of Repeated Exposure on Recognition, Familiarity, and Liking.' *Psychological Bulletin* 143 (5): 459–98; doi.org/10.1037/bul0000085.

Morewedge, Carey K., Haewon Yoon, Irene Scopelliti, Carl W. Symborski, James H. Korris, and Karim S. Kassam. 2015. 'Debiasing Decisions: Improved Decision Making with a Single Training Intervention'. *Policy Insights from the Behavioral and Brain Sciences* 2 (1): 129–40; doi. org/10.1177/2372732215600886.

Murase, Yohsuke and Christian Hilbe. 2024. 'Computational Evolution of Social Norms in Well-Mixed and Group-Structured Populations'. *Proceedings of the National Academy of Sciences* 121 (33): e240688512; doi.org/10.1073/pnas.2406885121.

NBC News. 2024. 'National Exit Polls: Election 2024 Results'. 20 November 2024; nbcnews.com/politics/2024-elections/exit-polls.

Neal, David T., Wendy Wood, Mengju Wu and David Kurlander. 2011. 'The Pull of the Past: When Do Habits Persist Despite Conflict With Motives?' *Personality and Social Psychology Bulletin* 37 (11): 1428–37; doi.org/10.1177/0146167211419863.

Nemet, Gregory F., and Evan Johnson. 2010. 'Willingness to Pay for Climate Policy: A Review of Estimates'. *SSRN Electronic Journal*; doi.org/10.2139/ssrn.1626931.

Németh, Dezső, Teodóra Vékony, Gabor Orosz, Zoltan Sarnyai and Leor Zmigrod. 2024. 'The Interplay between Subcortical and Prefrontal Brain Structures in Shaping Ideological Belief Formation and Updating'. *Current Opinion in Behavioral Sciences* 57 : 101385; sciencedirect.com/science/article/pii/S2352154624000366.

Nevins, Michael. 2011. *Meanderings in New Jersey's Medical History*. iUniverse.

Nguyen, Hung Vu, Mai Thi Thu Le, Chuong Hong Pham, and Susie S. Cox. 2024. 'Happiness and Pro-Environmental Consumption Behaviors'. *Journal of Economics and Development* 26 (1): 36–49; emerald.com/insight/content/doi/10.1108/JED-07-2021-0116/full/html.

Nisa, Claudia F., Jocelyn J. Bélanger, Birga M. Schumpe, and Daiane G. Faller. 2019. 'Meta-analysis of Randomised Controlled Trials Testing Behavioural Interventions to Promote Household Action on Climate Change'. *Nature Communications* 10 (1): 4545; nature.com/articles/s41467-019-12457-2.

Nisbett, Richard E., Geoffrey T. Fong, Darrin R. Lehman, and Patricia W. Cheng. 1987. 'Teaching Reasoning' in *Science* 238.

Nolt, John. 2011. 'How Harmful are the Average American's Greenhouse Gas Emissions?' *Ethics, Policy and Environment* 14 (1): 3–10.

Noonan, Douglas S., Lin-Han Chiang Hsieh, and Daniel Matisoff. 2013. 'Spatial Effects in Energy-Efficient Residential HVAC Technology Adoption'. *Environment and Behavior* 45 (4): 476–503; doi.org/10.1177/0013916511421664.

———. 2015. 'Economic, Sociological, and Neighbor Dimensions of Energy Efficiency Adoption Behaviors: Evidence from the US Residential Heating and Air Conditioning Market'. *Energy Research & Social Science* 10 : 102–13; sciencedirect.com/science/article/pii/S2214629615300153.

Nuland, Sherwin B. 2003. *The Doctors' Plague: Germs, Childbed Fever, and the Strange Story ff Ignac Semmelweis.* New York: WW Norton.

O'Keefe, Anne Marie and Richard W. Pollay. 1996. 'Deadly Targeting of Women in Promoting Cigarettes'. *Journal of the American Medical Women's Association* 51 : 67–69; columbia.akadns.net/itc/hs/pubhealth/p9740/readings/okeefe.pdf.

Olbrisch, Miriam. 2024. 'ICILS-Studie: "40 Prozent

können wenig mehr als einen Link anklicken"' ['ICILS Study: "40 Per Cent Can Barely Do More than Click on a Link"']. *Der Spiegel*, 12 November 2024, Panorama Section; spiegel.de/panorama/bildung/digitalkompetenz-von-schuelern-40-prozent-koennen-wenig-mehr-als-einen-link-anklicken-a-7e6b991a-dc5e-4859-b99b-588e39a8b2c8.

Oliver, Thomas. 1987. *The Real Coke, the Real Story*. Reprint Edition. New York: Select Penguin.

Orlowski-Yang, Jeff, Dir. 2020. *The Social Dilemma*. Documentary drama. Exposure Labs, Argent Pictures, The Space Program.

Osterhammel, Jürgen. 2014. *The Transformation of the World: A Global History of the Nineteenth Century*. Princeton: Princeton University Press.

Otto, Ilona M., Jonathan F. Donges, Roger Cremades, Avit Bhowmik, Richard J. Hewitt, Wolfgang Lucht, Johan Rockström, Franziska Allerberger, Mark McCaffrey, and Sylvanus S.P. Doe. 2020. 'Social Tipping Dynamics for Stabilizing Earth's Climate by 2050'. *Proceedings of the National Academy of Sciences* 117 (5): 2354–65.

Pacheco, Julianna. 2012. 'The Social Contagion Model: Exploring the Role of Public Opinion on the Diffusion of Antismoking Legislation across the American States'. *The Journal of Politics* 74 (1): 187–202; doi.org/10.1017/S0022381611001241.

———. 2013. 'Attitudinal Policy Feedback and Public Opinion: The Impact of Smoking Bans on Attitudes towards Smokers, Second-hand Smoke, and Antismoking Policies'. *Public Opinion Quarterly* 77 (3): 714–34; doi.org/10.1093/poq/nft027.

Pajula, Tiina, Saija Vatanen, Kaisa Grönman, and Laura

Lakanen. 2021. *Carbon Handprint Guide*. Espoo: VTT Technical Research Centre of Finland Ltd.

Palm, Alvar. 2017. 'Peer Effects pn Residential Solar Photovoltaics Adoption—A Mixed Methods Study of Swedish Users'. *Energy Research & Social Science* 26 : 1–10; sciencedirect.com/science/article/pii/S2214629617300087.

Palzer, Andreas and Hans-Martin Henning. 2014. 'A Comprehensive Model for the German Electricity and Heat Sector in a Future Energy System with a Dominant Contribution from Renewable Energy Technologies — Part II : Results'. *Renewable and Sustainable Energy Reviews* 30 (February): 1019–34; doi.org/10.1016/j.rser.2013.11.032.

Pandya, Apurvakumar and Pragya Lodha. 2021. 'Social Connectedness, Excessive Screen Time during Covid-19 and Mental Health: A Review of Current Evidence'. *Frontiers in Human Dynamics* 3: 684137; frontiersin.org/articles/10.3389/fhumd.2021.684137/full?ref=superhuman-blog.

Panksepp, Jaak. 1998. *Affective Neuroscience: The Foundations of Human and Animal Emotions*. New York: Oxford University Press.

Pendergrast, Mark. 2000. *For God, Country, and Coca-Cola*. New York: Basic Books.

Pfündel, Katrin, Anja Stichs, and Kerstin Tanis. 2021. *Muslimisches Leben in Deutschland 2020* [*Muslim Life in Germany 2020*]. Study commissioned by the German Islam Conference. Vol. 38. Study Report / Federal Office for Migration and Refugees (BAMF) — Migration, Integration and Asylum Research Centre (FZ).

Plato. 2021. *Meno*. Trans by Eva Brann, Peter Kalkavage,

and Eric Salem. Massachusetts: Hackett Publishing.

Poore, J. and T. Nemecek. 2018. 'Reducing Food's Environmental Impacts through Producers and Consumers'. *Science* 360 (6392): 987–92; doi.org/10.1126/science.aaq0216.

Porter, Patrick G. 1971. 'Advertising in the Early Cigarette Industry: W. Duke, Sons & Company of Durham'. *The North Carolina Historical Review* 48 (1): 31–43; jstor.org/stable/23518222.

Pronin, Emily, Thomas Gilovich, and Lee Ross. 2004. 'Objectivity in the Eye of the Beholder: Divergent Perceptions of Bias in Self Versus Others'. *Psychological Review* 111 (3): 781; psycnet.apa.org/record/2004-15929-009.

Prooijen, Jan-Willem van, Karen M. Douglas, and Clara De Inocencio. 2018. 'Connecting the Dots: Illusory Pattern Perception Predicts Belief in Conspiracies and the Supernatural'. *European Journal of Social Psychology* 48: 320–35.

Przybylski, Andrew K. and Netta Weinstein. 2017. 'A Large-Scale Test of the Goldilocks Hypothesis: Quantifying the Relations Between Digital-Screen Use and the Mental Well-Being of Adolescents'. *Psychological Science* 28 (2): 204–15; doi.org/10.1177/0956797616678438.

Quinn, Jeffrey M., Anthony Pascoe, Wendy Wood, and David T. Neal. 2010. 'Can't Control Yourself? Monitor Those Bad Habits'. *Personality and Social Psychology Bulletin* 36 (4): 499–511; doi.org/10.1177/0146167209360665.

Rajecki, D. W. 1974. 'Effects of Prenatal Exposure to Auditory or Visual Stimulation on Postnatal Distress

Vocalizations in Chicks'. *Behavioral Biology* 11 (4): 525–36; sciencedirect.com/science/article/abs/pii/S0091677374908451.

Rapeli, Lauri and Vesa Koskimaa. 2022. 'Concerned and Willing to Pay? Comparing Policymaker and Citizen Attitudes towards Climate Change'. *Environmental Politics* 31 (3): 542–51.

Reber, Rolf, Piotr Winkielman, and Norbert Schwarz. 1998. 'Effects of Perceptual Fluency on Affective Judgments'. *Psychological Science* 9 (1): 45–48; doi.org/10.1111/1467-9280.00008.

Rogers, Rebecca M., Cornelia Wallner, Bernhard Goodwin, Werner Heitland, Wolfgang W. Weisser, and Hans-Bernd Brosius. 2017. 'Analyzing The Existence and Relation of Optimistic Bias and First-Person Perception for an Impersonal Environmental Change'. *International Journal of Communication* 11: 20; ijoc.org/index.php/ijoc/article/view/5611.

Rollwage, Max, Raymond J. Dolan, and Stephen M. Fleming. 2018. 'Metacognitive Failure as a Feature of Those Holding Radical Beliefs'. *Current Biology* 28 (24): 4014–21; sciencedirect.com/science/article/pii/S0960982218314209.

Rothman, Alexander J., Roger D. Bartels, Jhon Wlaschin, and Peter Salovey. 2006. 'The Strategic Use of Gain- and Loss-Framed Messages to Promote Healthy Behavior: How Theory Can Inform Practice'. *Journal of Communication* 56 (suppl. 1): S202–20; academic.oup.com/joc/article-abstract/56/suppl_1/S202/4102613.

Rothman, Alexander J. and Peter Salovey. 1997. 'Shaping perceptions to motivate healthy behavior: The role of message framing'. *Psychological Bulletin* 121 (1): 3;

psycnet.apa.org/fulltext/1997-02112-001.html.

Russell, Stuart J. 2020. *Human Compatible: Artificial Intelligence and the Problem of Control*. London: Penguin Books.

Saint-Exupery, Antoine de. 2000. *Citadelle*. Paris: Gallimard-Jeunesse.

Sapolsky, Robert. 1994. *Why Zebras Don't Get Ulcers*. New York: Holt Paperbacks.

Sarason, Irwin G. and Charles D. Spielberger. 1975. *Stress and Anxiety: II*. Washington: Hemisphere Publishing.

Sauer, Hanno. 2024. *The Invention of Good and Evil: A World History of Morality*. London: Profile Books.

Schmitt, Michael T., Lara B. Aknin, Jonn Axsen, and Rachael L. Shwom. 2018. 'Unpacking the Relationships Between Pro-environmental Behavior, Life Satisfaction, and Perceived Ecological Threat'. *Ecological Economics* 143 : 130–40; tarjomanic.ir/wp-content/uploads/2022/06/Ecological-Economics-tarjomanic.pdf.

Schulte, Ulrich. 2021. 'Ulf Poschardt zur Mobilitätswende: ›Teslas sind öde Autos‹' ['Ulf Poschardt on the Mobility Revolution: "Teslas Are Bleak Cars"']. *taz*, 2 July 2021, Politics Section; taz.de/!5779417.

Schultz, P. Wesley, Taciano L. Milfont, Randie C. Chance, Giuseppe Tronu, Sílvia Luís, Kaori Ando, Faiz Rasool et al. 2014. 'Cross-Cultural Evidence for Spatial Bias in Beliefs About the Severity of Environmental Problems'. *Environment and Behavior* 46 (3): 267–302; doi.org/10.1177/0013916512458579.

Sellier, Anne-Laure, Irene Scopelliti, and Carey K. Morewedge. 2019. 'Debiasing Training Improves Decision Making in the Field'. *Psychological Science* 30 (9): 1371–79; doi.org/10.1177/0956797619861429.

Sharot, Tali. 2011. 'The Optimism Bias'. *Current Biology* 21 (23): R941–45; doi.org/10.1016/j.cub.2011.10.030.

——. 2012. *The Optimism Bias. A Tour of the Irrationally Positive Brain.* New York: Knopf Doubleday.

——. 2017. *The Influential Mind: What the Brain Reveals About Our Power to Change Others.* Little, Brown.

Sharot, Tali, Christoph W. Korn, and Raymond J. Dolan. 2011. 'How Unrealistic Optimism Is Maintained in the Face ff Reality'. *Nature Neuroscience* 14 (11): 1475–79; doi.org/10.1038/nn.2949.

Stables, Gloria J., Amy F. Subar, Blossom H. Patterson, Kevin Dodd, Jerianne Heimendinger, Mary Ann S. Van Duyn, and Linda Nebeling. 2002. 'Changes in Vegetable and Fruit Consumption and Awareness among US Adults: Results of the 1991 and 1997 5 A Day for Better Health Program Surveys'. *Journal of the American Dietetic Association* 102 (6): 809–17; pubmed.ncbi.nlm.nih.gov/12067046.

Stephens, John L. 1854. *Incidents of Travel in Central America, Chiapas, and Yucatan.* London: A. Hall, Virtue & Company.

Strauss, Delphine. 2023. 'Generative AI Set to Affect 300mn Jobs across Major Economies'. *Financial Times*, 27 March 2023, section: Artificial Intelligence; ft.com/content/7dec4483-ad34-4007-bb3a-7ac925643999.

Tainter, Joseph. 1988. *The Collapse of Complex Societies.* Cambridge: Cambridge University Press.

Trotsky, Lev Davidovich. 1933. *The Only Road.* Pioneer Publishers; marxists.org/deutsch/archiv/trotzki/1932/09/04-21fehler.htm.

Twenge, Jean M. 2019. 'More Time on Technology, Less Happiness? Associations Between Digital-

Media Use and Psychological Well-Being'. *Current Directions in Psychological Science* 28 (4): 372–79; doi.org/10.1177/0963721419838244.

Twenge, Jean M. and W. Keith Campbell. 2018. 'Associations Between Screen Time and Lower Psychological Well-being among Children and Adolescents: Evidence from a Population-based Study'. *Preventive Medicine Reports* 12 : 271–83; sciencedirect.com/science/article/pii/S2211335518301827.

Tyson, Alec and Shiva Maniam. 2016. 'Behind Trump's Victory: Divisions by Race, Gender, Education'. *Pew Research Center* (blog). 9 November 2016; pewresearch.org/short-reads/2016/11/09/behind-trumps-victory-divisions-by-race-gender-education.

Uddin, Lucina Q. 2021. 'Cognitive and Behavioural Flexibility: Neural Mechanisms and Clinical Considerations'. *Nature Reviews Neuroscience* 22 (3): 167–79; doi.org/10.1038/s41583-021-00428-w.

U.S. Department of Health and Human Services. 2000. 'Chapter 2 — A Historical Review of Efforts to Reduce Smoking in the United States'. In *Reducing Tobacco Use: A Report of the Surgeon General*. Atlanta volume. U.S. Department of Health and Human Services, Centers for Disease Control and Prevention.

Villegas, Paulina and Hannah Knowles. 2021. 'Iceland Tested a 4-Day Workweek. Employees Were Productive — and Happier, Researchers Say.' *Washington Post*, 7 July 2021; washingtonpost.com/business/2021/07/06/iceland-four-day-work-week.

Vlasceanu, Madalina, Kimberly C. Doell, Joseph B. Bak-Coleman, Boryana Todorova, Michael M. Berkebile-Weinberg, Samantha J. Grayson, Yash Patel, et al. 2024.

'Addressing Climate Change with Behavioral Science: A Global Intervention Tournament in 63 Countries'. *Science Advances* 10 (6): eadj5778; doi.org/10.1126/sciadv.adj5778.

Vogel, Jefim and Jason Hickel. 2023. 'Is Green Growth Happening? An Empirical Analysis of Achieved versus Paris-Compliant CO 2– GDP Decoupling in High-Income Countries'. *The Lancet Planetary Health* 7 (9): e759–69; doi.org/10.1016/S2542-5196(23)00174–2.

Vohra, Karn, Alina Vodonos, Joel Schwartz, Eloise A. Marais, Melissa P. Sulprizio, and Loretta J. Mickley. 2021. 'Global Mortality from Outdoor Fine Particle Pollution Generated by Fossil Fuel Combustion: Results from GEOS-Chem'. *Environmental Research* 195 (April): 110754; doi.org/10.1016/j.envres.2021.110754.

Wagner, Ulrich, Oliver Christ, Thomas F. Pettigrew, Jost Stellmacher, and Carina Wolf. 2006. 'Prejudice and Minority Proportion: Contact Instead of Threat Effects'. *Social Psychology Quarterly* 69 (4): 380–90; doi.org/10.1177/019027250606900406.

Watts, Nick, Markus Amann, Nigel Arnell, Sonja Ayeb-Karlsson, Jessica Beagley, Kristine Belesova, Maxwell Boykoff et al. 2021. 'The 2020 Report of The Lancet Countdown on Health and Climate Change: Responding to Converging Crises'. *The Lancet* 397 (10269): 129–70; doi.org/10.1016/S0140-6736(20)32290-X.

Weber, Hannes. 2019. 'Attitudes Towards Minorities in Times of High Immigration: A Panel Study among Young Adults in Germany'. *European Sociological Review* 35 (2): 239–57; academic.oup.com/esr/article-abstract/35/2/239/5258083.

Weiss, Harvey. 2017. *Megadrought, Collapse, and Causality. Megadrought and Collapse from Early Agriculture to Angkor*, 1–31.

Whittlesea, Bruce WA. 1993. 'Illusions of Familiarity'. *Journal of Experimental Psychology: Learning, Memory, and Cognition* 19 (6): 1235; psycnet.apa.org/record/1994-24230-001.

Winkielman, Piotr and John T. Cacioppo. 2001. 'Mind at Ease Puts a Smile on the Face: Psychophysiological Evidence that Processing Facilitation Elicits Positive Affect.' *Journal of Personality and Social Psychology* 81 (6): 989–1000; doi.org/10.1037/0022-3514.81.6.989.

Wittgenstein, Ludwig. 2009. *Philosophical Investigations*. 4th edition. Chichester: Wiley-Blackwell.

Wolf, Zachary B., Curt Merrill, Way Mullery, and CNN. 2024. 'Anatomy of Three Trump Elections: How Americans Voted in 2024 vs. 2020 and 2016'. 2024; cnn.com/interactive/2024/politics/2020-2016-exit-polls-2024-dg.

Wolf-Pommrich, Hedda. 1956. *Knigge-Brevier: ABC der Umgangsformen für Beruf und tägliches Leben* [*The Knigge Guide: An A-B-C of Good Manners for Professional and Everyday Life*]. Gabler Verlag.

Wood, Wendy. 2017. 'Habit in Personality and Social Psychology'. *Personality and Social Psychology Review* 21 (4): 389–403; doi.org/10.1177/1088868317720362.

Wood, Wendy and Dennis Rünger. 2016. 'Psychology of Habit'. *Annual Review of Psychology* 67 (1): 289–314; doi.org/10.1146/annurev-psych-122414-033417.

Wurthmann, L. Constantin. 2022. 'Empirische Analyse gesellschaftlicher Wertorientierungen in der Bundesrepublik Deutschland von 2009 bis 2017'. In

Wertorientierungen und Wahlverhalten ['An Empirical Analysis of Social Value Orientation in the Federal Republic of Germany 2009 to 2017' in *Value Orientation and Voter Behaviour*], 217–395. Springer VS, Wiesbaden; doi.org/10.1007/978-3-658-38456-2_4.

WZB [The Berlin Social Science Center]. 2013. *Datenreport 2013-Ein Sozialbericht für die Bundesrepublik Deutschland.* [*Data Report 2013 — A Social Report for the Federal Republic of Germany.*]

Yin, Henry H. and Barbara J. Knowlton. 2006. 'The Role of the Basal Ganglia in Habit Formation'. *Nature Reviews Neuroscience* 7 (6): 464–76; doi.org/10.1038/nrn1919.

Yon, Daniel, Cecilia Heyes and Clare Press. 2020. 'Beliefs and Desires in the Predictive Brain'. *Nature Communications* 11 (1): 4404; nature.com/articles/s41467-020-18332-9.

Yoon, Haewon, Irene Scopelliti, and Carey K. Morewedge. 2021. 'Decision Making Can Be Improved through Observational Learning'. *Organizational Behavior and Human Decision Processes* 162 (January):155–88; doi.org/10.1016/j.obhdp.2020.10.011.

Zajonc, Robert B. 1968. 'Attitudinal effects of Mere Exposure.' *Journal of Personality and Social Psychology* 9 (2p2): 1; psycnet.apa.org/journals/psp/9/2p2/1.

Zawadzki, Stephanie Johnson, Linda Steg, and Thijs Bouman. 2020. 'Meta-analytic Evidence for a Robust and Positive Association between Individuals' Pro-environmental Behaviors and Their Subjective Wellbeing'. *Environmental Research Letters* 15 (12): 123007; iopscience.iop.org/article/10.1088/1748-9326/abc4ae/meta.

Zmigrod, Leor. 2022. 'Individual-Level Cognitive and

Personality Predictors of Ideological Worldviews: The Psychological Profiles of Political, Nationalistic, Dogmatic, Religious, and Extreme Believers'. *PsyArXiv*; doi.org/10.31234/osf.io/srgup.

Zmigrod, Leor, Ian W. Eisenberg, Patrick G. Bissett, Trevor W. Robbins, and Russell A. Poldrack. 2021. 'The Cognitive and Perceptual Correlates of Ideological Attitudes: A Data-driven Approach'. *Philosophical Transactions of the Royal Society B: Biological Sciences* 376 (1822): 20200424; doi.org/10.1098/rstb.2020.0424.

Zmigrod, Leor, Peter J. Rentfrow, and Trevor W. Robbins. 2018. 'Cognitive Underpinnings of Nationalistic Ideology in the cCntext of Brexit'. *Proceedings of the National Academy of Sciences* 115 (19): E4532–40.

Zmigrod, Leor and Manos Tsakiris. 2021. 'Computational and Neurocognitive Approaches to the Political Brain: Key Insights and Future Avenues for Political Neuroscience'. *Philosophical Transactions of the Royal Society B: Biological Sciences* 376 (1822): 20200130; doi.org/10.1098/rstb.2020.0130.